The Reliability, Availability and Productiveness of Systems

The Reliability, Availability and Productiveness of Systems

D.J. Sherwin

Consultant
formerly School of Mechanical and Manufacturing Engineering
Queensland University of Technology
Australia

and

A. Bossche

Department of Electrical Engineering
Delft University of Technology
The Netherlands

SPRINGER-SCIENCE+BUSINESS MEDIA, B.V.

First edition 1993

Originally published by Chapman & Hall
Typeset in 10/12 Times by Best-set Typesetter Ltd., Hong Kong

ISBN 978-94-010-4688-6

A catalogue record for this book is available from the British Library

Library of Congress Cataloging-in-Publication data
Sherwin, D.J.
The Reliability, Availability, and Productiveness of Systems / D.J. Sherwin and A. Bossche. – 1st ed.
p. cm.
Includes bibliographical references and index.
ISBN 978-94-010-4688-6 ISBN 978-94-011-1582-7 (eBook)
DOI 10.1007/978-94-011-1582-7
1. Systems engineering. I. Bossche, A. II. Title.
TA168.S442 1993
620'.0045 – dc20 93-16765
CIP

Printed on permanent acid-free text paper, manufactured in accordance with the proposed ANSI/NISO Z 39.48-199X and ANSI Z 39.48-1984

Contents

Preface

This book is about the measurement and prediction of the reliability behaviour of systems of physical items. It is not specifically concerned with human factors with safety analysis as such, although some of the techniques discussed are adaptable to these purposes.

A machine or an electronic circuit exemplifies a system. Each machine or circuit may also be treated as an item in a larger system. However, this does not reduce it suddenly to basic component status; it remains complex and can only be treated as unitary under definable restrictions. In particular, the effects of maintenance and component renewal must be considered most carefully.

Previous books on system reliability have concentrated on one or two only of the six principal techniques available to the analyst. These are:

1. probability theory;
2. distributional statistics;
3. markov methods (matrix algebra);
4. fault and event trees (Boolean logic);
5. theory of renewal processes;
6. directional graph theory (di-graphs).

This book relates these methods to one another and to their applications. The authors feel that previous books which concentrated upon one technique and the contortions necessary to use it in every possible situation may have misled readers into believing that there were no other methods and that some real problems were intractable or more difficult to solve than need be. For example, several results which are proved in other books for items with exponentially distributed times to/between failures are shown to be independent of distribution.

Both authors are engineers rather than statisticians and take a practitioner's view of the degree of rigour required to 'prove' mathematical results. The aim is to make each technique and its limitations clear to working engineers rather than to satisfy the pure mathematician with pedantic proofs which would remain opaque to target readers and distract attention from the practical uses of the results. Every effort has therefore been made to keep the mathematical explanations as simple as possible. Some prior knowledge of algebra, calculus, probability and statistics is assumed, to about the level usually reached in European courses for bachelor's degrees in engineering.

Acknowledgement

Extracts from BS5760: Part 1: 1985 are reproduced with the permission of BSI. Complete copies of the standard can be obtained by post from BSI Publications, Linford Wood, Milton Keynes, MK14 6LE.

ERRATA

page 66, equation (5.20) should read:

$$\frac{1}{(1 + a^2)}\begin{bmatrix} \cdots & \cdots & \cdots \\ 2a^2 & 2a(1 - a) & (1 - a)^2 \\ \cdots & \cdots & \cdots \end{bmatrix}$$

page 80, equation (6.23) should read:

$$\phi = r(n - m + 2 - r)/\mu + \sum_{i=1}^{r-1} i/\mu$$

page 170, line 8 should read: 3. Now the edges e_4, e_5 and e_7 together with vertex v_4 can be reduced to ...

page 184, line 11 should read: ... draw two lines of slope $1/S$ with equations ...

page 228, lines 2–4 should read: ... example at Fig. 13.8. This form is designed to be placed to the right of the corresponding FMEA form at Fig. 13.6.

page 236, line 12 should read: $k \geq 1$, k repairmen,

page 239, line above Fig. 13.10 should read: ... making Weibull plots of $(s^2 + t^2)^{½}$ against mean or median rank of ...

The Reliablity, Availability and Productiveness of Systems.
D.J. Sherwin and A. Bossche. Published in 1993 by Chapman & Hall, London. ISBN 0 412 39320 4.

CHAPTER 1

Definitions, terminology and basics

1.1 DEFINITIONS OF TERMS AND MEANINGS OF SYMBOLS

In this book, the definitions of terms used conform generally to those given in the British Standards BS4778 'Glossary of terms used in quality assurance (including reliability and maintainability terms)', BS5532 'Statistics – Vocabulary and symbols', and BS3811 'Glossary of maintenance terms in terotechnology'.

Readers are assumed to have some previous acquaintance with the concepts of probability and statistics at an elementary level. Those that have never studied these subjects before should read the early chapters of *Statistics for Technology* by Christopher Chatfield (1983), or some other such primer.

A complete list of symbols used in this book appears in Appendix C. They are generally the same as those used in the British Standard BS5760. Each new symbol will be explained when first introduced. Here follows a 'starting set'.

t	Age at failure of a single component.
T	Age of a system containing fallible components.
$f(t)$	Probability density function of ages at failure of a component (p.d.f.).
$F(t)$	Cumulative distribution function of $f(t)$ (c.d.f.).
$R(t)$	Reliability function, $1 - F(t)$.
$z(t)$	Hazard rate function, $f(t)/R(t)$.
$H(t)$	Cumulative hazard function, integral of $z(t)$ from 0 to t
$P(X)$	Probability of event X.
t_r	Time to repair a single component or mode of failure.
$g(t_r)$	p.d.f. of repair times.
$M(t_r)$	Maintainability, c.d.f. of repair times.
θ	Mean of $f(t)$; mean time to failure; MTTF of component; or MTBF of a system
ϕ	Mean of $g(t)$; mean time to repair, MTTR.
A	Steady state availability; MTT(B)F/[MTT(B)F + MTTR].

$A(t)$ Probability item is available at time t, given available at time 0, called time-dependent availability.

λ Constant hazard rate (exponential distribution), or constant rate of occurrence of failures (ROCOF) for a system.

$\lambda(T)$ ROCOF in a repairable system, which varies with system age T.

The concept of reliability generally envisages only one life and one failure, whereas that of availability includes repairs as needed and periods of available or 'up' time alternating with periods of repair or 'down' time following failures. Although both concepts apply both to single items and to systems, generally availability is more useful for systems whereas the separate concepts of reliability and maintainability are more useful for single items. This is because most systems are repairable. In fact, if they are not repairable, then one might as well regard them as single items for most purposes.

The remainder of this section on definitions and meanings is devoted to showing how the various terms in the starter set are related to each other.

Most readers will be familiar with the concept of a distributed variable, often expressed in terms of its mean and variance or mean and standard deviation. Books on elementary statistics tend to dwell upon the normal or Gaussian distribution, but we shall see later that in reliability work it is the exception rather than the rule. The following observations are applicable to any distribution form of a continuous nature.

The p.d.f. is best understood by stating that the elemental area under the curve between t and $t + \mathrm{d}t$ represents the probability of failure in that short interval. This area is algebraically $f(t)\mathrm{d}t$. The integral of this expression from t_1 to t_2 is the unconditional probability that an item taken at random from a population having failure time p.d.f. $f(t)$ fails between those two times. The case where t_1 is zero is of particular interest. This is the probability that an item fails before t_2, represented as $F(t_2)$. If on the other hand, t_1 is positive and t_2 is positive infinite, the resulting area represents the probability that failure is delayed until after t_1. This is the same as the probability of survival from zero to t_1, and is called the reliability to t_1, $R(t_1)$. The generalized functions associated with these probabilities are called the cumulative distribution function or c.d.f., $F(t)$, and the reliability or survivor function, $R(t)$. Clearly from these definitions, and referring to Fig. 1.1

$$R(t) \equiv 1 - F(t), \tag{1.1}$$

where

$$F(t) \equiv \int_0^t f(u)\mathrm{d}u. \tag{1.2}$$

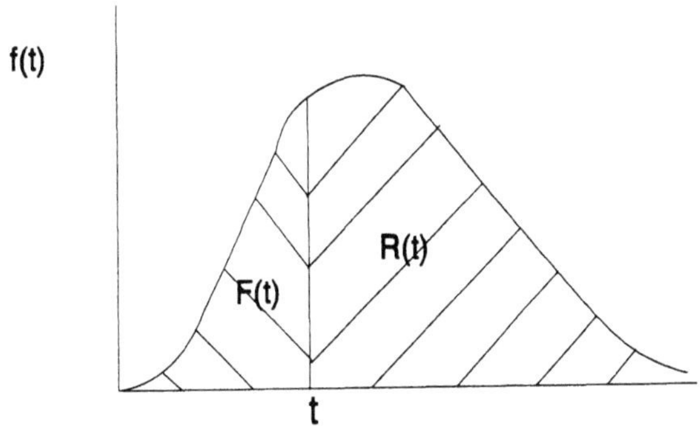

Fig. 1.1 Relationship of p.d.f., c.d.f. and reliability function for a continuous distribution of times to failure.

We shall also be much concerned with the instantaneous rate of failures occurring to those items of a population which have survived to time t, which is denoted by $z(t)$ and is called the hazard rate. We have already seen that the unconditional probability of failure in the interval t to $t + \mathrm{d}t$ is $f(t)\mathrm{d}t$, so the unconditional rate of failure at time t is simply $f(t)$, since $\mathrm{d}t$ is infinitesimal. 'Probability density' and 'rate' are synonymous when the variable is time. The probability of survival to t is $R(t)$. So, to obtain the conditional rate of failure of survivors to t we must divide $f(t)$ by $R(t)$,

$$z(t) \equiv f(t)/R(t). \tag{1.3}$$

There is one more function associated with the distribution form with which we shall later be concerned. This is the cumulative hazard function, $H(t)$, which is the integral from zero to t of $z(t)$ (Fig. 1.2)

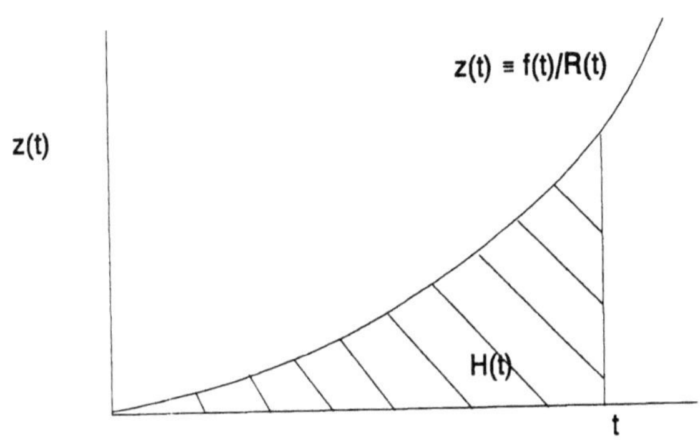

Fig. 1.2 Hazard rate function $z(t)$ derived from Fig. 1.1 showing cumulative hazard function as an integral of $z(t)$.

$$H(t) \equiv \int_0^t z(u)\mathrm{d}u. \tag{1.4}$$

Using the defining relationships between the various functions of a distribution, several other relationships which apply to any continuous distribution may be derived by simple algebra and calculus as follows:

$$f(t) = -\mathrm{d}R(t)/\mathrm{d}t \tag{1.5}$$

$$z(t) = -\mathrm{d}\log_e R(t)/\mathrm{d}t \tag{1.6}$$

$$R(t) = \exp[-H(t)]. \tag{1.7}$$

It is important for the reader to understand that these relationships apply to any continuous distribution form, including empirical ones. Any curve whose total integral is unity can be used as a p.d.f.

Maintainability is defined as the probability that a repair or renewal of an item is completed within a time t_r. It is therefore the cumulative distribution function of the distribution of repair times evaluated at t_r

$$M(t_r) \equiv \int_0^{t_r} g(x)\mathrm{d}x. \tag{1.8}$$

The *mean* of a distribution is the first moment of area of the p.d.f. about the origin over the whole range, which is usually 0 to ∞ for failure and repair time distributions. Negative times are meaningless in reliability work. It is usually understood that an item cannot fail before it is started for the first time, and a negative repair time is obviously nonsense

$$\theta = \int_0^{\infty} tf(t)\mathrm{d}t \tag{1.9}$$

$$\phi = \int_0^{\infty} t_r g(t_r)\mathrm{d}t_r. \tag{1.10}$$

The mean is the average or expected value of the failure or repair time.

1.2 REVIEW OF PROBABILITY THEORY – BAYES' THEOREM

It is prudent to review the basic tenets of probability theory because calculations of system reliability and availability from the figures for items in the system depend upon it. We expect an unbiased coin to return 50% heads and 50% tails, and we say that the probability of heads on any one throw is 0.5. But if we test 10 diesel engines for 5000 h each or until failure whichever occurs first we do not know what number of failures to expect. If two engines fail we might estimate $R(5000)$ as about 0.8, but we would prefer to have more data. The probability required could only be established by an infinite number of tests. $R(5000)$ is then the limit as the number of trials tends to infinity of the ratio of successes (s) to trials (n).

$$R = \underset{n \to \infty}{Lt}\,(s/n) \tag{1.11}$$

The number of failures r is always $n - s$. This is equivalent to saying that $R = 1 - F$, where F is the failure probability as n tends to infinity. Some books on reliability, and also BS4778 consider a 'true' reliability to be established when all members of the population have failed. However, if the figure is to be of any use to anyone there must be more examples of the item extant and unfailed. In practice therefore, all reliability-related probabilities are estimates from samples of true values that could only be established by an infinite and therefore impossible number of experiments!

A principal aim of this book is to instruct the reader how to estimate probabilities for complex systems from data or estimates relating to the system's components. This requires not only knowledge of the theory of simple probability, but also that of conditional probability and Bayesian concepts.

If an event becomes more or less likely to occur if another event has occurred already, then the other event is said to condition the first. The conditional probability of event A given event B is written $P(\mathrm{A}|\mathrm{B})$. If A and B are independent events then $P(\mathrm{A}|\mathrm{B}) = P(\mathrm{A})$, but otherwise they are different. For example, $P(\mathrm{A})$ might be the probability that an item fails before 1000 h regardless of where it is fitted. Event B might then be the case in which the item is fitted in an environment likely to affect its reliability, so that $P(\mathrm{A}|\mathrm{B})$ is less than the average unreliability $P(\mathrm{A})$. Bayes' theorem is concerned with deducing the probability of B given A from knowledge of $P(\mathrm{A})$, $P(\mathrm{B})$, and $P(\mathrm{A}|\mathrm{B})$. In our example, $P(\mathrm{B}|\mathrm{A})$ represents the probability that the item was in the failure-inducing environment given the event of its failure. $P(\mathrm{B})$ is the unconditional probability that it was fitted in the failure-inducing environment which can be estimated as the proportion of the total number of items fitted which are disadvantaged in this way. Bayes' theorem is stated as follows

$$P(\mathrm{B}|\mathrm{A}) = P(\mathrm{A}|\mathrm{B})P(\mathrm{B})/P(\mathrm{A}). \tag{1.12}$$

Note that $P(\mathrm{B}|\mathrm{A})$ is the proportion of item failures coming from the failure-inducing environment, which is taken to be the same as the probability that a failed item of unknown history was so disadvantaged.

EXAMPLE 1.1

A factory making pistons acquires gudgeon pins from sources A and B in proportions 25:75. The gudgeon pins have reliabilities 0.999 and 0.9999, respectively. Find the overall probability of failure and the probability that a failed gudgeon pin of uncertain origin is in fact from source A.

SOLUTION

The overall probability of failure of the gudgeon pins is

$$P(\mathrm{F}) = 0.25 \times (1 - 0.999) + 0.75 \times (1 - 0.9999) = 0.000\,325.$$

By Bayes' theorem

$$P(\mathrm{A}|\mathrm{F}) = P(\mathrm{F}|\mathrm{A})P(\mathrm{A})/P(\mathrm{F}),$$

where $P(\mathrm{A}|\mathrm{F})$ is the probability we require

$$P(\mathrm{A}|\mathrm{F}) = 0.001 \times 0.25/0.000\,325 = 0.7692.$$

Note that $P(\mathrm{A})$ is called the *prior* probability that the source was A and $P(\mathrm{A}|\mathrm{F})$ is the *posterior* probability given that it had failed, that the item is from source A.

Statisticians divide into two camps over Bayesian matters. Probabilists insist that history does not affect the probabilities of the various possible outcomes of the next trial whereas Bayesians say that it does. For instance, a probabilist would say that the probability of a die coming up six at the next trial is always 1/6 even if the last three trials had all been sixes. A Bayesian would consider whether there was yet enough evidence to pronounce the die biased. In more general terms, the Bayesian has various ways of weighting the prior probabilities and the observations to arrive at a posterior estimate. For example, giving equal weight to the prior and experimental estimates would result in a posterior probability of $(1/6 + 3/3)/2 = 0.5833$ that the next throw would be six. Ultimately all methods of setting the relative weights are subjective. Note that the probabilist and the Bayesian would agree if they both were convinced that the die was not biased. The Bayesians would argue as follows, applying their hero's theorem

$$\begin{aligned} P(\{4 \times 6\mathrm{s}\}|\{3 \times 6\mathrm{s}\}) &= P(4 \times 6\mathrm{s})P(\{3 \times 6\mathrm{s}\}|\{4 \times 6\mathrm{s}\})/P(3 \times 6\mathrm{s}) \\ &= (1/6)^4 1/(1/6)^3 = 1/6. \end{aligned}$$

Both groups would reject the idea that a fourth six is most unlikely given three in a row, i.e. $(1/6)^4$ just because the die was known to be fair. Only inveterate gamblers hold this view; it is equivalent to the mistaken belief that their luck is bound to change soon in a game in which the probabilities are fixed by the physics of the apparatus, as in dice.

System reliability theory, however, normally relies on combining *independent* probabilities. We have considered conditional probability first only to emphasize that where independence is not present, matters do become more complicated. The basic rules for independent probabilities are very simple indeed. A stricter and more extensive exposition of these rules is given in Chapter 8; those given here are sufficient for the present. Readers may find it helpful to refer to the Venn diagram (Fig. 1.3).

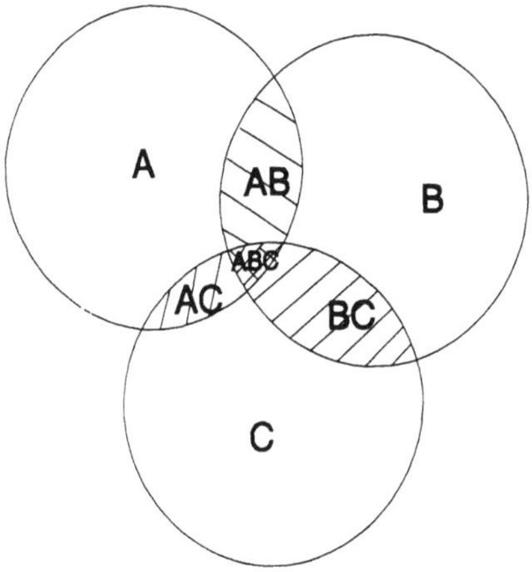

Fig. 1.3 Venn diagram illustrating combinations of three probabilities.

$$P(\text{A and B and C} \ldots) = P(\text{A})P(\text{B})P(\text{C}). \ldots \quad (1.13)$$

$$\begin{aligned} &P(\text{A or B or C or any combination}) \\ &\quad = P(\text{A}) + P(\text{B}) + P(\text{C}) - P(\text{A})P(\text{B}) - P(\text{B})P(\text{C}) \\ &\quad\quad - P(\text{A})P(\text{C}) + P(\text{A})P(\text{B})P(\text{C}). \end{aligned} \quad (1.14)$$

$$\begin{aligned} &P(\text{A or B or C but not combinations}) \\ &\quad = P(\text{A}) + P(\text{B}) + P(\text{C}) - 2P(\text{A})P(\text{B}) - 2P(\text{A})P(\text{C}) \\ &\quad\quad - 2P(\text{B})P(\text{C}) + 3P(\text{A})P(\text{B})P(\text{C}). \end{aligned} \quad (1.15)$$

$$P(\text{none of A,B,C,} \ldots) = [1 - P(\text{A})][1 - P(\text{B})][1 - P(\text{C})]. \ldots \quad (1.16)$$

$$P(\text{A but not B}) = P(\text{A})[1 - P(\text{B})]. \quad (1.17)$$

In the case of equation (1.14) the triple overlap ABC appears as part of each of the first three terms, i.e. three times. Then when we take away the double overlaps AB, BC and AC it is subtracted three times, so finally it must be put back once. In equation (1.15) we wish to subtract all overlaps. The double terms must be subtracted twice to remove them altogether. The triplet is added three times in the single probabilities, subtracted six times in the double terms and so must be added back three times in the final term. The obvious extensions of (1.14) and (1.15) for four and more items are valid.

If A and B are such that B cannot occur without A, then by Bayes' theorem

$$\begin{aligned} P(\text{B}\,|\,\text{A}) &= P(\text{B})/P(\text{A}) \\ \text{since } P(\text{A}\,|\,\text{B}) &= 1. \end{aligned} \quad (1.18)$$

1.3 DISTRIBUTION FORMS USED IN RELIABILITY

A distribution form is simply a family of formulae, which, interpreted graphically, always produce a total integral over their full ranges of exactly one. This is of course a working rather than a formal definition. The formulae are known as probability density functions or probability mass functions, p.d.f.s or p.m.f.s. Readers are assumed to be already familiar with p.d.f.s. Probability mass functions apply to discrete distribution forms such as the binomial and the Poisson in which the variable can take only integer values.

This section is intended as both introduction and for future reference. It lists the characteristics of the principal forms of distribution used in reliability. Readers are also recommended to study the very useful little book on the subject by Hastings and Peacock (1980).

1.3.1 Binomial distribution

The probability of exactly x failures out of a possible number of items at risk n is binomially distributed by definition. Let the probability of failure of each independent, identical item be p. Then the p.m.f. is given by

$$P(x/n) = \binom{n}{x} p^x (1 - p)^{n-x}, \qquad (1.19)$$

where $\binom{n}{x} = {}^nC_x = n!/[x!(n - x)!]$ which is the number of different ways that exactly x items out of n can be failed.

The distribution function is simply the sum of such terms from 0 to x. The binomial distribution is useful in reliability in determining the reliability of systems of identical items in which some minimum number $m \geqslant 1$ must be unfailed for system success. More generally, it is useful in problems where the total number of items is fixed and limited and they exist in one or other of only two states.

The mean of the Binomial is np and the variance is $np(1 - p)$ both of which may be fractional. The modal or most likely value is the integer in the range $p(n + 1) - 1$ to $p(n + 1)$.

EXAMPLE 1.2

Five items were placed on test for 5000 h to see if their specified reliability to that time of 0.8 had been met. Three items failed to reach 5000 h. Should we accept the design for mass production on the basis of this test?

SOLUTION

The probability of exactly three failures out of five where $p = 0.2$ is

$$\binom{5}{3} 0.2^3 0.8^2 = 0.0512.$$

Adding on the probabilities of four and of five failures when $p = 0.2$ brings the total probability of a result as bad or worse than the one we got to 0.0579. Thus we can reject the hypothesis that the design meets the specified reliability with over 90%, but less than 95% confidence. Alternatively, the same arithmetic gives us the probability of failure of a three-out-of-five system of items of reliability 0.8.

Tables of the binomial distribution listing the probability of x or more events out of n when the individual probability is p are readily available.

1.3.2 Poisson distribution

In this case we are concerned with the probability of x events when the average number is m. The number of events is *not* limited as it was in the binomial case. In reliability work the mean m is often the product of a rate of occurrence of failures or constant hazard rate and a time of operation. It is a limiting form of the binomial as $n \to \infty$ and $p \to 0$. In these circumstances

$$n!p^x/(n-x)! \to (np)^x \quad \text{and} \quad (1-p)^{n-x} \to \exp(-np).$$

This gives us for the p.m.f.

$$f(x) = m^x \exp(-m)/x!. \tag{1.20}$$

Statistical texts state that the Poisson distribution describes the variation in numbers of isolated, random events in equal periods in a continuum of time. Failures to a complex maintained system in which only what has failed is renewed fit this pattern empirically, but only failures caused by truly random events such as freaks of the weather conform theoretically to a true Poisson distribution.

EXAMPLE 1.3

It is known that the average ROCOF of a power system is 35 failures in 10 000 h. In one particular year an example of this design suffers only one failure. Find the probability of this event, and that of no failures at all in a year.

SOLUTION

The mean number of failures in a year is first found. There are $24 \times 365.25 = 8766$ h in an average year, so the expected number m is given by

$$m = 35 \times 8766/10\,000 = 30.681.$$

Probability of exactly one failure is

$$f(1) = 30.681 \times \exp(-30.681)/1 = 1.453 \times 10^{-12}$$

which is a most unlikely event prompting investigation; perhaps it was not switched on for most of the year. No failures at all is even less likely at

$$f(0) = 1 \times \exp(-30.681)/1 = 4.736 \times 10^{-14}$$

noting that $m^0 \equiv 1$ and $0! \equiv 1$.

As with the binomial distribution, it is usually the probability of x or more events with which practitioners are concerned. Although in this example the probability of zero failures is two orders smaller than that of one failure, this will not always be so. In practice, this problem would have been solved by looking up the probability of two or more failures when $m = 30$ and 31 and interpolating if thought to be worthwhile. Only the more voluminous tables list $m = 31$, so it is as well to remember that a normal approximation is adequate for $m > 9$ using the fact that the variance of a Poisson distribution is equal to its mean. Thus the distribution in this case is approximately normal with mean 30.681 and s.d. $30.681^{\frac{1}{2}} = 5.539$.

$$\mu = \frac{30.681 - 1}{5.539} = 5.36 \text{ s.d.s.}$$

From tables the area in the tail of the distribution is less than 10^{-9}. Although in this case the calculation is two or three orders in error the accuracy is nevertheless sufficient to make the right decision. Also, when there are none or one or two events, calculation from first principles is feasible. If the number of failures had been, say, 10 the normal approximation would be very much the quicker, i.e.

$$\mu = \frac{30.681 - 10}{5.539} = 3.7 \text{ s.d.s.}$$

$$\text{Area in tail} = 0.000\,11.$$

We would still be suspicious even with 10 failures that special factors applied in this case.

The probability of zero failures is also the reliability function of the exponential distribution with constant hazard rate m/T, where T is the time over which m is the expected number of failures, in our example 8766 h for 30.681 failures or 10 000 h for 35 failures, giving a hazard rate of 0.0035/h. But in our example we do not have a constant hazard rate but a constant ROCOF arising from a maintained system. This is explained in more detail in the section on the bathtub curve. Suffice it to say here that the arithmetic remains the same for the ROCOF case but the calculation represents the *conditional* reliability over a period of duration T given that the system was operational at the start of the period and that no changes were made to the maintenance policy during that period. Note that the Poisson distribution represents purely random behaviour, characterized by a complete lack of 'memory'. A failure is no more or less likely in any particular equal period; the probability is not influenced at all by how long it may have been since the previous failure. The mathematics of the Poisson distribution follow from these assumptions.

1.3.3 Exponential distribution

The exponential distribution is very popular in reliability work because it has only one parameter, the mean or its reciprocal the constant hazard rate. Referring back to equations (1.1) to (1.7), given a constant hazard rate as the defining feature of the distribution, we can infer the other functions thus

$$z(t) = \text{constant, say } \lambda.$$

Then
$$R(t) = \exp(-\lambda t) \tag{1.21}$$

and
$$f(t) = \lambda \exp(-\lambda t). \tag{1.22}$$

As mentioned above, the reliability function is also the probability of no failures in the corresponding Poisson distribution with $m = \lambda t$. The exponential and the Poisson distributions are in fact different ways of expressing the same statistical situation, which is the operation of pure chance with no 'memory'. Most engineers would instinctively incline to the view that hazard rate is an increasing function of time for most mechanical components, and they would be right on the evidence of collected data. The exponential distribution is inappropriate where things wear out or fatigue or corrode away. It is commonly used (possibly strictly misused) for electronic components. These have very low hazard rates anyway and often become obsolete before they get old enough for the shape of the curve to matter practically.

1.3.4 Weibull distribution

The Weibull is the most commonly used distribution form in modern reliability engineering practice. This is because it is very versatile, being able to handle monotonically increasing or decreasing hazard rates. The exponential distribution is a special case of the Weibull distribution. Its basis is that the hazard rate function is taken to be a power function of time. However, so that the parameters can be either dimensionless or have the dimension of time only, the actual parameterization seems at first arcane. It has two or three parameters as follows:

1. a is the scale parameter or characteristic life often written η (eta).
2. b is the shape parameter (dimensionless), often written β (beta).
3. c is a location parameter, minimum time to failure, often written γ (gamma) and often zero.

Then

$$f(t) = (b/a)[(t - c)/a]^{b-1} \exp[-\{(t - c)/a\}^b] \tag{1.23}$$

$$R(t) = \exp[-\{(t - c)/a\}^b] \tag{1.24}$$

$$z(t) = b(t - c)^{b-1}/a^b \tag{1.25}$$

$$\theta = a\Gamma(1 + 1/b) + c \tag{1.26}$$

$$\sigma^2 = a^2\Gamma(1 + 2/b) - (\theta - c)^2. \tag{1.27}$$

When b is 1 and $c = 0$ the Weibull distribution reduces to the exponential distribution, when $b > 1$ the hazard rate increases with t and when $b < 1$ it falls with t. The Weibull distribution is therefore the starting assumption of distribution form in most analyses of field data. Special graph paper is available upon which the distribution function plots as a straight line (for $c = 0$). Even when the distribution form which best fits the data is not Weibull, such a plot will suggest to the experienced eye which other distribution forms might fit better. This is explained in Chapter 12.

1.3.5 Normal distribution or Gaussian distribution

The p.d.f. of the normal distribution is

$$f(t) = (1/\sigma\sqrt{2\pi}) \exp[-(t - \theta)^2/2\sigma^2]. \tag{1.28}$$

The reliability and hazard rate functions do not exist in closed form but the distribution function is extensively tabulated. The normal distribution is not much used in reliability work because it starts at $-\infty$, and of course failures before $t = 0$ are not usually considered possible. The advantage is that the mean and s.d. are the parameters θ and σ. Where the mean is three or more times the s.d., this restriction is not important, and in this connection the normal distribution is used to represent failure times by abrasive wear, which it does quite well. If the nearest Weibull distribution has $b > 3$ the normal distribution may well fit just as well or even better.

1.3.6 Lognormal distribution

If the logarithms rather than the failure times themselves are normally distributed, we have a lognormal distribution

$$f(t) = (1/ts\sqrt{2\pi}) \exp[-(\log t/m)^2/2s^2]. \tag{1.29}$$

The parameter m is the distribution median, and s is the s.d. of the logarithms of the failure or repair times. This distribution is the only one known to the authors with as few as two parameters in which the hazard rate rises to a maximum and then falls again to an asymptotic value. Although Wallodi Weibull originally developed his distribution for fatigue experiments, such failures often exhibit this form of hazard rate function and so the lognormal may be a better fit. Another use for the lognormal distribution is as the distribution of repair times. There are good theoretical reasons to expect a lognormal distribution when the task is the sum of several successive elements each randomly distributed, such as finding a book in a library. Much maintenance work consists of eliminatory diagnosis of faults, successively narrowing the list of possible causes after which the actual repair is relatively straightforward. The mean and s.d. are

$$\theta = m\exp(s^2/2),\ \sigma^2 = m^2\exp(s^2)[\exp(s^2) - 1],$$
$$\text{mode} = m/\exp(s^2/2). \tag{1.30}$$

1.3.7 Gamma distribution

The gamma distribution was used as the standard form before it was superseded by the Weibull. Like the Weibull distribution, it can represent falling, constant or rising hazard rates, as the shape parameter, c, moves from fractional to unity to greater than one. It still has importance in the system reliability field as the distribution of times to failure of 1-out-of-n standby systems. It represents the sum of c failure times, each identically and exponentially distributed with hazard rate z.

$$f(t) = z^c t^{c-1}\exp(-zt)/\Gamma c, \tag{1.31}$$

where

$$\Gamma c = \int_0^\infty \exp(-u)u^{c-1}du.$$

Note that $\Gamma c = (c - 1)!$ if c is an integer in which case the distribution form is called the Erlang, having been developed by him to describe queues in telephone systems.

The mean and variance are

$$\theta = c/z,\ \sigma^2 = c/z^2 \tag{1.32}$$

As c increases the distribution tends to the normal. Mathematically it is the c-fold convolution of the exponential with hazard rate z. This means that two or more distributions of gamma form and scale parameter z can be added to form another gamma in which their shape parameters are added, i.e.

$$[G\colon z,c_1] + [G\colon z,c_2] = [G\colon z,(c_1 + c_2)] \tag{1.33}$$

Also, if $2c$ is integer and $z = 1$ then the gamma distribution is related to the chi-square distribution, which is tabulated, as follows

$$[G\colon 1,c] = [\chi^2\colon 2c]/2. \tag{1.34}$$

1.3.8 Extreme value distribution or Gumbel distribution

The extreme value distribution was proposed by Gumbel as a model of the distribution of extreme heights of floods to assist in the design of flood defences. In reliability work it is usually the distribution of the smallest extreme values that is of interest, for example the time for the first of an unknown number of corrosion sites in a pipe to progress to a hole. In the Gumbel distribution the hazard rate function is of exponential form, so that the logarithms have a Weibull distribution. The distribution has two parameters, the mode a, and a scale parameter b. The distribution of the smallest extreme is negatively skewed (mean less than mode), a shape that the Weibull and other two-parameter distributions cannot cover. For positive skew (largest extreme), it is usually adequate and always easier to use the Weibull model.

$$f(t) = (1/b)\exp[(t-a)/b][\exp\{-\exp(t-a)/b\}] \tag{1.35}$$

$$R(t) = \exp[-\exp\{(t-a)/b\}] \tag{1.36}$$

$$z(t) = (1/b)\exp[(t-a)/b] \tag{1.37}$$

$$\theta = a - 0.577\,21b,\ \sigma^2 = b^2\pi/6. \tag{1.38}$$

1.3.9 Hyperexponential distribution

Hyperexponential means simply beyond the exponential, in reliability terms it is on the 'other side' of exponential from the wear-out forms of distribution. Mathematically it implies a distribution in which the s.d. exceeds the mean. If $b < 1$ the Weibull is hyperexponential and this form is often used, but this implies an infinite mode at zero, which is unrealistic. A more realistic form is the weighted sum of two exponentials, and the simplest parameterization in that family is

$$f(t) = 2k^2\lambda\exp(-2k\lambda t) + 2\lambda(1-k)^2\exp[-2(1-k)\lambda t] \tag{1.39}$$

$$R(t) = k\exp(-2k\lambda t) + (1-k)\exp[-2(1-k)\lambda t] \tag{1.40}$$

$$z(t) = f(t)/R(t) \tag{1.41}$$

with $0 < k \leqslant 0.5$ throughout.

1.3.10 Beta distribution

The beta distribution is a continuous inversion of the binomial distribution. It represents the likelihood function of the binomial probability p in terms of two shape parameters v,w. In reliability, this is the likelihood function of the value of the failure time distribution at the time of the ith ordered failure out of a sample of n, where $v = i$ and $w = n - i + 1$. It is the basis of the distribution estimation method most commonly used with failure data, i.e. median ranks. In this case the 50% or median point of the likelihood function for each failure is used as the estimate of $F(t)$ to plot against the time of failure. Mean ranks are also used with larger samples.

$$f(p) = p^{v-1}(1 - p)^{w-1}/B(v,w), \tag{1.42}$$

where $B(v,w)$ is the beta function.

$$B(v,w)= \Gamma v\Gamma w/\Gamma(v + w).$$

$$\theta= v/(v + w) \quad \text{or} \quad i/(n + 1) \tag{1.44}$$

for $v = i$, $w = n - i + 1$.

$$\sigma^2 = vw/(v + w)^2(v + w + 1)$$
$$\text{or} \quad i(n - i + 1)/(n + 1)^2(n + 2). \tag{1.45}$$

$$\text{Mode} = (v - 1)/(v + w - 2)$$
$$\text{or} \quad (i - 1)/(n - 1)v,w > 1. \tag{1.46}$$

$$\text{Median estimate (Bénard)} = (v - 0.3)/(w + v - 0.6)$$
$$\text{or} \quad (i - 0.3)/(n + 0.4). \tag{1.47}$$

Note that this distribution also can be skewed either way, positive for $v < w$ and negative for $v > w$. It is available in tabulated form, but reliability engineers usually use tables of median ranks and also 5% and 95% ranks to plot the distribution estimate and its confidence limits. Bénard's estimate is very accurate for $n > 5$.

A four-parameter version of the beta distribution is also used. Whereas the two-parameter version runs from 0 to 1, the four-parameter version runs from a to $a + b$. So the range or scale is b and the location of the start is at $+a$ from zero. The p.d.f. is obtained by substituting $(t - a)/b$ for p and $(a + b - t)/b$ for $1 - p$ in the two-parameter version, thus

$$f(t) = [(t - a)/b]^{v-1}[(a + b - t)/b]^{w-1}/B(v,w) \tag{1.48}$$

$$\theta = a + bv/(v + w),\ \sigma^2 = b^2vw/(v + w)^2(v + w + 1). \tag{1.49}$$

$$\text{Mode} = a + b(v - 1)/(v + w - 2). \tag{1.50}$$

The four-parameter beta distribution is used in project management to estimate the mean times for activities in critical-path analysis. It is a property of the distribution that

$$\theta = [a + 4m + (a + b)]/6, \tag{1.51}$$

where m is the mode, so one simply estimates longest, shortest and most likely times, combines them as above and obtains an estimate of mean time for the activity. The same approach can be used to estimate failure mean times in the absence of recorded data, by asking maintainers and operators for the three estimates.

1.4 THE BATHTUB CURVES FOR COMPONENTS AND SYSTEMS

1.4.1 Introduction

The bathtub curve is perhaps the oldest and best-known concept in reliability engineering. It is also the cause of a great deal of confusion and woolly thinking. The two different concepts are of hazard rate for single components which are renewed (or restored to good as new) after each failure on the one hand and the overall rate of occurrence of failures (ROCOF) of a system in which components are renewed or restored as they fail and such that after a while all fallible components have different ages. It is confusion between these two concepts that is at the root of the bathtub curve muddle. There are in fact two distinct bathtub curve concepts (Fig. 1.4), one for single components which plots $z(t)$ versus t, and the other for complex systems which plots ROCOF versus system age T. It is legitimate to use the $z(t)$ concept for complex items only if every failure of any component is followed by either renewal or overhaul to good-as-new of all fallible components. Thus, for example, a printed circuit board which is renewed as a whole on failure, may be treated as a component with a combined $z(t)$ curve consisting of the sum of the hazard rates of all its components. On the other hand, a pump consisting of casing, shaft impeller, sealing rings, bearings and mechanical seal is definitely a system in which components are renewed as they fail. After some service, all the components have different ages and the overall rate of occurrence of failures (ROCOF), $\lambda(T)$, after T hours of operation since the whole pump was brand new depends upon those ages, and their individual hazard rate functions

$$\lambda(T) = \sum_i z_i(t_i), \tag{1.52}$$

where t_i is the current age of component i and $z_i(t_i)$ its hazard function.

1.4.2 The hazard rate (component) bathtub curve

The $z(t)$ bathtub is said to arise as follows. Populations of new components always contain a proportion of defectives which fail early. This gives the $z(t)$ curve an initial downward trend, levelling out as t gets larger. The constant $z(t)$ portion of the curve which follows is explained by a combination of random failures due to occasional overloads, with the tail of the early failures and the start of wearout failures. The final, rising portion of the curve is ascribed to wearout becoming the dominant mode of failure. The curve is often said to apply to electronic components. Whilst there is no doubt that mass-produced parts are almost bound to contain some defectives, the recent advances in manufacturing and quality

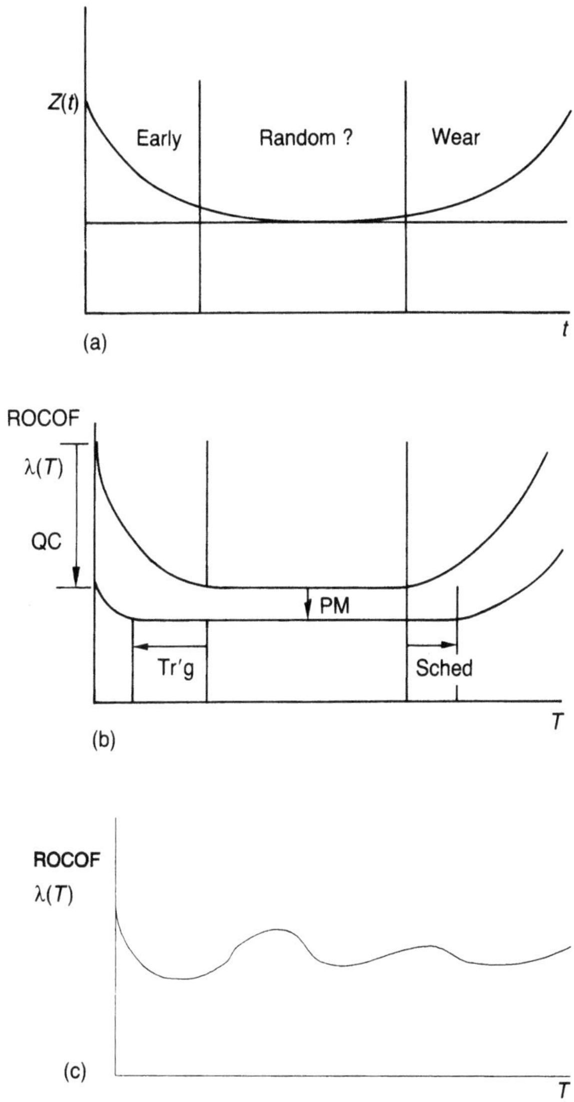

Fig. 1.4 (a) Component bathtub curve of hazard rate. (b) System bathtub curves showing effects upon system ROCOF of QC, training of operators and maintainers, PM intervals and extending the PM schedule. (c) Wave pattern generated by individual system realization of bathtub theory.

control have eliminated most of the problem. As it is known that temperature, shock and vibration all enhance the recorded hazard rates of electronic components, it follows that the mechanisms of failure are probably mechanical in nature and they are therefore likely to be increasing rather than constant. Data-banks such as MIL-HDBK 217 usually give higher average rates than actually experienced, possibly due to a combination of the effects of lower load due to redundancy, and because the data were compiled from service records in which uncertainty existed as to which component failed first and which others were blown by the resulting overload, all being debited with a failure from a single incident. Most electronic components never reach the true wearout phase before they are scrapped. A lot of later failures may be due to stresses imposed during burn-in, a procedure in which new circuits are tested at over-stress conditions to make the quality failures occur in the factory rather than the field. It is at least arguable that components are not always dichotomously divided into good and defective, some may have the defect to a lesser degree and fail in the middle of their so-called useful lives due to the well-intentioned ill-treatment received during burn-in and other over-stress testing. This is certainly the case with mechanical components, which usually start to wear out as soon as they start work, that is $z(t)$ starts at $t = 0$ and rises monotonically.

Defective mechanical components are more often obvious and do not usually get fitted, but if they do occur then there are usually only two modes, not three phases, to the failure pattern. Mechanical components are often virtually failure-free, suggesting that design is important, particularly the application of materials science at an early stage. Some mechanical components are deliberately designed to fail in preference to others, for example drive shafts in vehicles should break before the studs otherwise a wheel falls off; these are inevitably subject to more service failures than the parts they protect.

1.4.3 The ROCOF (system) bathtub curve

The $\lambda(T)$ or ROCOF bathtub has not only a different timescale, but also a different plotted variable. It has only its shape and the fact that it is concerned with reliability in common with the $z(t)$ bathtub. The timescale is system time from new, T, rather than component age t. On this T timescale, components within the system which have been renewed will have ages less than T. On the T scale, after an initial transient, the combined ROCOF consists of the instantaneous sum of all the $z_i(t_i)$ values for the fallible components, each measured from the component's last renewal, as in equation (1.52). For a multi-component system this produces a ROCOF curve with 'waves' of random length and amplitude, the mean of which tends to be constant provided that the maintenance and operational policies are not changed. Because the waves are random and particular to each replication of the system, all the past history and all the hazard rates would have to be known to make even a short-term prediction other than the average value. All long-term predictions would be of the average value. So, although in any individual manifestation of the system design there would be waves, for a population of systems and in predictions, the best estimate is a constant ROCOF predicated upon a constant operating and maintenance policy. The initial transient mentioned above may be falling or rising. It is usually falling and this is due to a combination of quality and assembly faults, installation errors, design errors, operator errors and maintenance errors which do not recur. As the poor quality components are renewed with good ones, the design and installation errors corrected, and the operators and maintainers gain experience, the ROCOF falls to its constant average value. It does not start to rise again consistently until long-lifed items start to fail for the first time. It is when this phase is apparent that renewal or overhaul should be considered. During the so-called 'useful life' period between the teething problems and the final phase, the value of $\lambda(T)$ is constant if averaged over several examples of the system. In the absence of preventive maintenance and redundancy (duplication of components) this average value will be the sum of the reciprocals of the individual component distribution means, i.e.

$$\lambda(T) = \sum_{i=1}^{n} 1/\theta_i, \tag{1.53}$$

where there are n components which are repairable or renewable.

1.4.4 Discussion – human mortality

The shape of the $z(t)$ bathtub cannot be altered much, except to reduce the initially high value by better quality control. The rest of the curve is regarded as characteristic of the component design as it is stressed in the application under consideration, and virtually unalterable. But the shape of the ROCOF curve is quite readily altered by the application of various measures in varying degrees. All these measures cost money. For a price the initial problems can be reduced. In well-established designs which show the results of years of expensive experience, such as centrifugal pumps, the ROCOF curve may even start from zero and approach its constant value from below. In new unevolved designs, total quality management, operator and maintainer training, and installation by makers followed by acceptance trials can much reduce the potential level and duration of the initial transient. But matching pumps to applications means paying skilled and experienced personnel to provide a service to customers, and this puts up the pumps' prices. The measures for new designs, which may include extensive reliability programs are obviously expensive, although it is often cheaper to pay their cost than to suffer unreliable equipment later. The constant value of ROCOF can be reduced by preventive maintenance. This also costs money, but may be worthwhile when the failures saved and their attendant delays are considered. When the curve turns upward, it is usually the case that the failures become more expensive as well as more frequent. The question then arises as to whether the system should be renewed, replaced by a better system or overhauled to extend its life. If the last course is chosen, on grounds of cost again, then after the overhaul, a new bathtub curve will be started. If the overhaul is to 'good as new' then the new curve will have, in theory, the same characteristics as the original, but in practice it is usually shorter and with a higher constant ROCOF value.

In summary, the shape of the ROCOF bathtub is not immutable, it can be changed by policy decisions particularly those concerning design procedures, quality control in manufacture, and maintenance and overhaul in service. The overall objective is usually to minimize the mean life-cycle costs per unit time of ownership and operation. Figure 1.5 shows how these decisions may affect the total cost curve, and the economic life of the system over a single bathtub. Note that the result of particular policy changes may be to increase or decrease total life mean cost rate and it is not always easy to predict which will occur if the system is being operated near-optimally to start with.

Human mortality is often quoted as an example of a practical bathtub curve. It is also one of the few cases in which there is no shortage of data.

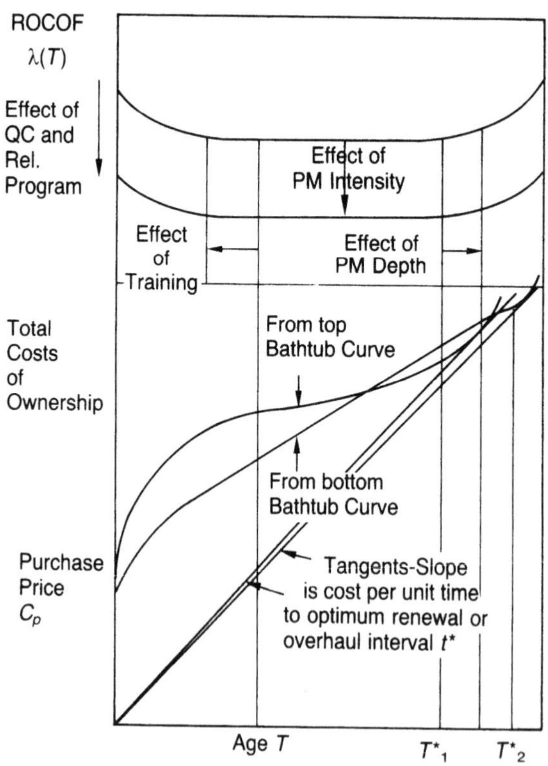

Fig. 1.5 Effect of QC training, PM intensity and PM depth on bathtub curves, costs and durability of complex items.

Examination of mortality curves however, shows no flat portion, but rather a single minimum around the early thirties. Distributional analysis gives a bimodal rather than a tri-modal model. This is not surprising, since the human body is all of the same age (apart from isolated instances of spare-part surgery) and dies when the first vital organ fails. The early deaths are due to congenital weaknesses, accidents of youth, etc., but it seems that accidental death is never a sufficiently major cause to dominate and produce a constant mid-portion to the curve. It is also noteworthy that if one survives to about 80 the curve peaks and falls quite sharply, perhaps reflecting society's care for the old and frail. It is interesting to speculate upon what may happen to these curves in the future when people are persuaded to take better care of their bodies. Will a constant phase appear or not? Are exercise and healthier diet analogous to maintenance? This example has probably added to the past confusion; the human body is not unequivocally either a maintained system or one in which all the $z_i(t)$s have the same time-base and can legitimately be added, although it is perhaps closer to the latter than some previous writers have realized.

1.5 AVAILABILITY

Availability is a combination of reliability and maintainability. Every repairable or renewable item starts 100% available at $t = 0$, but the probability that it is available at some instant t from new is a function of both $f(t)$ and $g(t_r)$. The more remote t is from zero, the less it matters whether the item was available or under repair at $t = 0$. After a while this instantaneous availability $A(t)$ settles down to the steady-state value A given by

$$A = \theta/(\theta + \phi). \tag{1.54}$$

A repairable system also has a steady-state availability defined in the same way as above but in terms of the system values of θ and ϕ, i.e. θ_{sys}, ϕ_{sys}. Thus

$$A_{sys} = \theta_{sys}/(\theta_{sys} + \phi_{sys}). \tag{1.55}$$

In both cases the verbal interpretation is that steady-state availability is the proportion of the total of operating and repair time spent in an operating condition.

For instantaneous and other time-related expressions of availability, mathematical formulae are obtainable only when both $f(t)$ and $g(t_r)$ are of exponential form. In other distributions $z(t)$ is not constant and so the values of reliability and maintainability depend upon age and so upon previous failure history. The instantaneous availability for exponential distributions with MTBF θ and MTTR ϕ is

$$A(t) = \frac{\theta}{\theta + \phi} + \frac{\phi}{\theta + \phi}\, e^{-(1/\theta + 1/\phi)t}. \tag{1.56}$$

Note that as $t \to \infty$ the second term of this expression disappears, confirming the steady-state result. $A(t)$ assumes that at $t = 0$ the item is in the 'up' or working state rather than the 'down' or under repair state and that there are no other states of the system that are relevant to the calculation. This expression is most easily proved by methods given in Chapter 5.

The mean availability of an item over a period T from a start in the up state is obtained by integrating $A(t)$ from 0 to T and dividing by T. This gives for exponential failure and repair time distributions

$$\bar{A}(T) = \frac{\theta}{\theta + \phi} + \frac{\theta^2\phi}{T(\theta + \phi)^2}\left[1 - e^{-(1/\theta + 1/\phi)T}\right]. \tag{1.57}$$

$\bar{A}(T)$ also tends to the steady-state value as T increases.

In most books these expressions are put in terms of $\lambda = 1/\theta$ and $\mu = 1/\phi$. We have written them in θ, ϕ because even those which strictly apply only to exponential distributions can usually be used quite safely as approximations for other distribution forms. The steady-state expression is distribution free, and the others degenerate to very close to A within $3(\theta + \phi)$.

In military circles it is often required to know what proportion of missions of duration T will be successful, where success is defined as freedom from failures taking longer then t_r to fix. This is called mission availability and is generally complicated to calculate because the durations T for which various items of a system are required are not all the same and may even be themselves distributed. However, in the simplest case of a single T, and for constant hazard rates, we may proceed as follows. Let the number of failures occurring in T be n. The probability that n takes a particular value j is Poisson, i.e.

$$P(n = j) = (T/\theta)^j e^{-T/\theta}/j!. \tag{1.58}$$

The probability that j out of j successive repairs will all be completed in t_r or less is

$$P(j \text{ out of } j) = (1 - e^{-t/\theta})^j. \tag{1.59}$$

The mission availability must cover all values of j from 0 to ∞, therefore

$$\begin{aligned} A_m &= \sum_{j=0}^{\infty}[P(n = j)P(j \text{ out of } j)] \\ &= \sum_{j=0}^{\infty}[(T/\theta)^j T e^{-T/\theta}(1 - e^{t/\phi})^j/j!] \\ &= \exp[-T\exp(-t_r/\phi)/\theta] = R(T)^{[1-M(t_r)]}. \end{aligned} \tag{1.60}$$

Note that in practice, not all missions start with all items in the up state. If this is so then steady-state values should be used. A_m values in general will be higher than steady-state values because of the assumption that items are up at the mission start.

If the item is up at the start of the mission and its age T_1 is known, then

$$A_m = [R(T_1) - R(T_1 + T)]^{[1-M(t_r)]}. \tag{1.61}$$

For constant $\lambda(T) = \lambda$ and constant $\phi_{sys} = 1/\mu$, equation (1.61) becomes

$$A_m = [\exp(-\lambda T)]^{[\exp(-\mu t_r)]}. \tag{1.61a}$$

We have considered so far only intrinsic availability expressions, that is those which consider only operating and active repair time in their calculation. This is because intrinsic values are, in our view, the only ones which are unlikely to change unless the item is redesigned or operated at different levels of stress. There is a plethora of other definitions and divisions of time to include logistical and managerial delays and to account for time when the item is in working condition but not actually required to run. We would only counsel readers to be very sure about what is being considered and recorded in this connection. We decline to add to the confusion of terms used in different ways by various authorities. We regard the only civilian terminology standard, BS4778, as inadequate in respect of availability.

REFERENCES

BS3811, *Glossary of maintenance terms in terotechnology*, British Standards Institution, London.

BS4778, *Glossary of terms used in quality assurance*, British Standards Institution, London.

BS5532, *Statistics – vocabulary and symbols*, British Standards Institution, London.

BS5760, *Reliability of systems, equipment and components*, British Standards Institution, London.

Chatfield, C. (1983) *Statistics for Technology*, Chapman and Hall, London.

Hastings, N.A.J. and Peacock, J.B. (1980) *Statistical Distribution*, Butterworth, Guildford.

MIL-HDBK 217 (regularly updated) *Reliability Prediction of Electronic Equipment*, RomAir Development Center, Griffis Air Force Base, NY 13441.

CHAPTER 2

Series systems

In reliability the words 'series' and 'parallel' are borrowed from electrical engineering. However, their meanings are changed. Items are in series for reliability only if failure of the system results from failure of any item. Two booster rockets might be employed to launch a missile. If both are needed for a successful launch then they are in series for reliability but in parallel for operation; if one rocket can fail without failing the mission then they are in parallel for both reliability and operation. With this proviso we can use block diagrams like electronic 'black box' diagrams to depict the relationships.

2.1 HISTORICAL – LÜSSER'S RULE

The history of series reliability, indeed that of reliability itself as an application of statistics to engineering, began in Hitler's Germany during World War II. An engineer named Lüsser was tasked to explain why early versions of the V1 or doodlebug, the world's first guided missile, often failed to reach their target, London. He surmised that without a pilot to make adjustments, any and every component failure would cause missile failure. The V1 was effectively a small unmanned aircraft with wings, fuselage and tail. On top was a simple pulse-jet engine. The fuselage contained a rudimentary guidance system to keep the missile on course, a timer to cut the engine over the target area, and of course a large explosive charge with percussion detonator. The pulse-jet was started by launching the missile up a ramp using a rocket. Each and every item on the following list of events had to occur as planned if the V1 was to achieve its objective of destroying one or two English homes.

1. Booster rocket fires correctly.
2. Pulse-jet starts.
3. Guidance system works for required time.
4. Pulse-jet works for required time.
5. Timer stops engine.
6. Detonator sets off main charge on impact.

None of these events is certain to occur every time; each has a probability of occurrence, say R_i, $i = 1, 2, \ldots, 6$. Some of the component reliabilities,

R_i, are time dependent and others are not. The probability of mission success for the missile is the intersection of all these events which is the product of the R_is. The general rule for calculating system reliability from component reliabilities is therefore a *product* rule.

$$R_{\text{system}} = \prod_{i=1}^{N} R_i. \tag{2.1}$$

This is also the rule for time-independent events and one-shot devices such as ammunition which have just one chance of working correctly and either do so or fail.

2.2 TIME-DEPENDENT RELIABILITY – NON-PARAMETRIC RESULTS

For the product rule to apply without error, the events or components concerned must be either sequential in time, as in the V1, or completely independent of one another. In most systems of components, the independence criterion rather than the sequential is the operative one. We are usually concerned with systems in which all the components operate continuously until one of them fails, but the example of the V1 illustrates that the product rule is more general.

Notice also that the times for which different components in a system must operate to achieve mission success may not be all the same. For example, in a warship the engines, navigational equipment and hotel services must operate to defined minimum levels throughout a mission, whereas the weapons systems may only be required for a few exciting minutes. Most items in the system therefore have, in the most general case, an appropriate operating time t_i, which is less than or equal to the mission time t, and a corresponding reliability term $R_i(t_i)$. The rest of the component reliabilities are simple time-independent probabilities. A more general expression for series system reliability is therefore

$$R_{\text{sys}}(t) = \prod_{i=1}^{N-M} R_i(t_i) \prod_{j=1}^{M} R_j, \tag{2.2}$$

where there are N items altogether of which M are one-shot devices.

The statistical distributions from which the functions $R_i(t_i)$ are derived (Chapter 1) need not be of the same form. The result at equation (2.2) is non-parametric or independent of distribution. In this book as many results as possible will be presented non-parametrically. Time-dependent and one-shot events can of course be combined as in the calculation for the V1 missile.

2.2.1 System mean time between failures, MTBF

Over a long period of operation, T, each of the components of a repairable system will on average fail T/θ_i times, where θ_i is the MTBF of item i. Each item failure is also a system failure, so the system MTBF is given by

$$\theta_{sys} = 1 \Big/ \sum_{i=1}^{N} [1/\theta_i]. \tag{2.3}$$

(Equation (1.53) expresses (2.3) in a different way.)

This result is often 'proved' in terms of constant failure rates and based on the exponential distribution. This is unnecessarily restrictive and can be misleading. Equation (2.3) is still true when the components are themselves complex. The rule applies to any series system however it may, for convenience, be split up. If item i is renewed rather than refurbished then substitute MTTF for MTBF.

2.2.2 Maintainability

Maintainability was defined in Chapter 1 as the probability that repair is completed within some arbitrary time t_r. Clearly, the system value of maintainability depends on the reliability and maintainability functions of all the N components of the system. Now, in general, the reliability functions vary with the times since that component last failed and all these times will, in general, be different. The system maintainability is therefore a function of all the times since last failure of all the components as well as of the 'allowed' repair time, t_r. Only for the first failure from new will all the ages of components be equal. In this special case the conditional probability that system failure is caused by item i rather than one of the other components is

$$F_i(t)/F_{sys}(t) = [1 - R_i(t)]/[1 - R_{sys}(t)]. \tag{2.4}$$

If item i causes the failure, the maintainability is $M_i(t_r)$. The system maintainability is therefore given by

$$M_{sys}(t_r,t) = \sum_{i=1}^{N} [F_i(t)M_i(t_r)]/F_{sys}(t). \tag{2.5}$$

Equation (2.5) is accurate and non-parametric but practically useless because

1. it applies only to the first failure;
2. it depends upon t as well as t_r.

As a general rule we shall give long-term or steady-state results where total generality is not possible. These arise when the system has been operating for a time long enough for all the fallible components of the system to have failed several times. The long-term, steady-state probability that a failure is caused by item i rather than one of the other items is given by θ_{sys}/θ_i which is the ratio of the item's mean rate of failure to that of the system. The system maintainability is then

$$M_{sys}(t_r) = \sum_{i=1}^{N} [M_i(t_r)/\theta_i]\,\theta_{sys}. \tag{2.6}$$

2.2.3 System mean time to repair, MTTR

The mean time to repair a series system is found in much the same way as the maintainability. Each component MTTR, ϕ_i must be weighted according to its relative frequency of failure θ_{sys}/θ_i. Thus

$$\phi_{sys} = \sum_{i=1}^{N} [\phi_i/\theta_i]\,\theta_{sys}. \tag{2.7}$$

2.2.4 Availability

In Chapter 1 various types of availability were defined for components having (unimodal) failure time distributions. In theory, all these types of availability are applicable to systems generally and series systems in particular. However, steady-state availability is by far the most useful form and the most easily calculated. Instantaneous and interval availabilities for systems are only calculable if all the component failure and repair time distributions are exponential in form.

(a) Steady-state availability

On average, each period of system uptime lasts θ_{sys}, and each succeeding period of downtime lasts ϕ_{sys}. As steady-state availability is defined as the long-term proportion of total time that the system is available or up, the simplest form for A_{sys} is given by

$$A_{sys} = \theta_{sys}/[\theta_{sys} + \phi_{sys}]. \tag{2.8}$$

In practical calculations based on equation (2.8), it is minimally necessary to know all the component MTBFs and MTTRs. We can then approach (2.8) through (2.3) and (2.7). But what if only the component steady-state availabilities A_i, $i = 1, 2, \ldots, N$, are known? It is tempting to suggest the product rule, but this is not accurate, as is explained in the next section. The product rule can, however, be used as an approximation when N is small, say less than 10, and all the A_i are large, say greater than 0.99. (Maximum error is less than 0.5%.) Even if just one of the component availabilities is small there will be little error, but if more than one is less than about 0.9 the error, which always leads to underestimation of A_{sys}, starts to become serious. It is particularly important not to

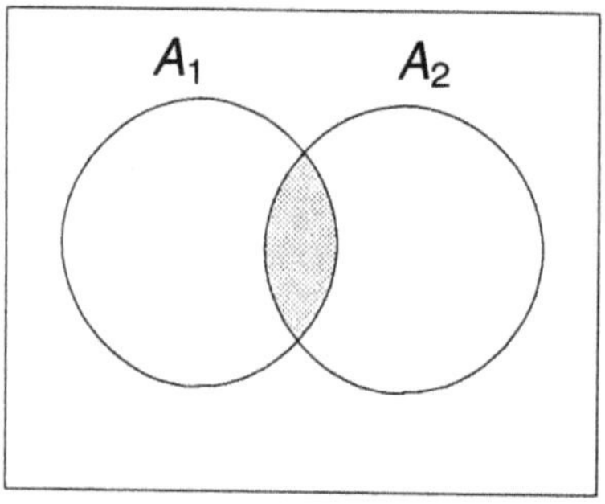

Fig. 2.1 Venn diagram representing the probability that both components are operating and the total area enclosed by curves (union) is the probability one or both of them is operable.

underestimate the availability of large series systems as this can lead to serious errors in estimating their economics, so the point is a vital one in industrial applications.

Consider a system of just two components in series. Let their independent steady-state availabilities be A_1 and A_2. Their situation can be represented on a Venn diagram (Fig. 2.1) as two overlapping circles having areas A_1 and A_2 within a square representing the total probability-space which has, of course, unit area. The overlapped area (intersection) then represents the probability that both components are operating and the total area enclosed by curved lines (union) is the probability that one or both of them is operable. The area of the intersection is given by the product of the component availabilities and the union is their sum less their product. The system steady-state availability is then

$$A_{1,2} = A_{\text{sys}} = A_1A_2/[A_1 + A_2 - A_1A_2]. \tag{2.9}$$

This result is independent of the forms of the failure and repair time distributions involved. Furthermore, it can be extended to systems with more than two components. If we add a third component to the system, we can treat items 1 and 2 as a single item and apply equation (2.9). This approach leads to a recurrence relation for the combined availability of the first i items in a series

$$A_{1,i} = A_{1,i-1}A_i/[A_{1,i-1} + A_i - A_{1,i-1}A_i]. \tag{2.10}$$

Equation (2.10) is tedious for the calculation of a long series but is easily coded for computer. It is easy to show by successive substitution in (2.10) that a general formula for combining series steady-state availabilities is

$$A_{\text{sys}} = 1\Bigg/\left[1 + \sum_{i=1}^{N}\{(1 - A_i)/A_i\}\right]. \tag{2.11}$$

This may be rearranged as follows

$$(1 - A_{sys})/A_{sys} = \phi_{sys}/\theta_{sys} = \sum_{i=1}^{N}[(1 - A_i)/A_i] = \sum_{i=1}^{N}(\phi_i/\theta_i). \quad (2.11a)$$

(b) *Why the product rule is wrong for availability*

Several previous work on system reliability have asserted the product rule for availability as well as reliability. But there is no doubt that it is wrong and that it can cause serious under-estimation of series availability. It is wrong because it does not take into account the fact that when any component is in the failed state, the others are not at risk of failure until it has been repaired. In terms of the Venn diagram at Fig. 2.1, the probability space is reduced from unity to the union of all the availabilities, that is, from 1 to $[A_1 + A_2 - A_1A_2]$. The product probability is therefore conditioned by the union probability.

The above analysis assumes that the availabilities of components are calculated from their actual running times and repair times. *The product rule is correct if component availabilities are based on total system time or calendar time*. The latter method of counting up and down times is however most unusual, and is only mentioned because it is not completely unknown; we have met it in the offshore oil industry. In this arcane interpretation item availability is redefined effectively as

$$a_i = (\text{total uptime of item } i)/(\text{total up} + \text{downtime of system}). \quad (2.12)$$

The total uptime of item i in this definition includes time when i could not operate because of failures elsewhere. The restraint that caused the product rule not to be true has therefore been removed.

Practitioners are therefore warned to be careful about the interpretation of data generally and of this point in particular.

(c) *Time-dependent availability*

We saw in Chapter 1 that there were also two forms of time-dependent availability, instantaneous availability (probability that the system is operational at some instant after starting from time-zero in known condition), and the average availability over a fixed period also with known *starting condition*. It was quite easy to calculate these availabilities for single components considered in isolation. However, in a system, even a simple series system, it becomes much more difficult. After one component has failed and been repaired it has a different age from all the others, and matters become progressively more complicated after each successive failure. As each component has its own failure and repair time distributions, it becomes impossibly tedious to find mathematical expressions and it is usual to resort to computer simulation to obtain estimates. It is, however pertinent to question whether such a simulation is likely to be worthwhile. This is because if T is large relative to system MTBF then the instantaneous, average and steady-state availabilities all tend to converge.

To deal effectively with the problem of finding either instantaneous or average time-dependent system availabilities, we need an estimate of the *system* failure time and repair time distributions in terms of the component distributions. We have seen already that the means of these distributions can be found (equations (2.3) and (2.7)). The most common practical problems involving time-dependent availability are concerned with systems which have been operated previously for a long period and therefore start each mission or period of required operation with all components in working order but all of randomly differing ages due to previous failures and repairs. In these mathematically special but practically common circumstances, it can be shown that the expected rate of occurrence of failures at time T from new tends to a constant value. System failure therefore becomes a pseudo-Poisson process. Although random oscillations about the constant average value appear for individual systems, if several simulations are overlaid and averaged the mean ROCOF always appears constant. Doubters may easily convince themselves of this by comparing the results of simulation over a random interval remote from system-new with those from the following equations.

The interval reliability of a series system remote from system-new and repaired by component renewal at and only at failure is approximately given by

$$R_{sys}(T_1,T_2) = \exp[-(T_2 - T_1)/\theta_{sys}], \tag{2.13}$$

where the interval runs from T_1 to T_2. This is the probability that the system suffers no failure between T_1 and T_2 *given* that there was no failure unrepaired at T_1.

Treating the failure of the system as a Poisson process, the probability of exactly j failures in T running hours is

$$P(j,T) = (T/\theta)^j \mathrm{e}^{-T/\theta}/j!. \tag{2.14}$$

The probability that a repair is completed in t_r or less is, by definition, the maintainability of the system. If follows that the probability that T running hours are completed with no failure taking longer than t_r to repair is

$$A_m(T,t_r) = \sum_{j=1}^{\infty}[P(j,T)\,M(t_r)^j], \tag{2.15}$$

where θ and $M(t_r)$ are system values. If repair can also be regarded as a pseudo-Poisson process then there is further simplification to

$$A_m(T,t_r) = \exp[-(T/\theta)\exp(-t_r/\phi)]. \tag{2.15a}$$

This version of time-dependent availability is often called the mission availability, but it does not, as has been claimed elsewhere, represent the probability that a mission of total duration T is not interrupted by failure for more than t_r on any single occasion. The mission duration, by equation (2.15), would average a little more than T but would always contain T operational hours.

For the case where both failure and repair can be regarded as Poisson processes, the instantaneous availability or probability that the system is operational at some instant t into a mission, having started out operational at time-zero, can be found by matrix methods detailed in Chapter 5. This gives

$$A(t) = \theta/(\theta + \phi) + [\phi/(\theta + \phi)]\exp\{-t/\theta - t/\phi\}. \qquad (2.16)$$

The first term of this expression is the steady-state availability. The second term is the decreasing time-dependent element at time t. At $t = 0$, $A(t) = 1$, and as t increases the second term becomes smaller and smaller. As we get away from zero and have more and more failures, the initial conditions have less and less influence upon the current availability. The result for Poisson processes suggests a more general result as follows

$$A(t) = A_{sys} + (1 - A_{sys})[1 - M_{sys}(t)]R_{sys}(t) \qquad (2.17)$$

but we have not been able to prove this mathematically. Equation (2.16) is therefore applicable to systems remote in time from new and having exponentially distributed repair times, which start each mission in a fully available condition.

The average availability over a mission of limited duration T may also occasionally be of interest. This can be found by integrating the expression for interval availability over 0 to T and then dividing by T, giving

$$\bar{A}(T) = \theta/(\theta + \phi) + [\theta\phi^2/T(\theta + \phi)^2][1 - \exp(-T/\theta - T/\phi)] \quad (2.18)$$

If the mission is a long one, the second term of equation (2.16) is very small. It consists of the product of the steady-state unavailability and the probability that a repair started at the beginning of the mission remains uncompleted at the end. These are both small relative to the steady-state availability. The third sub-term is the system reliability over the mission period, which also gets progressively smaller as the time increases. The decline of the second term is so rapid that it is not worth calculating if the mission period exceeds say twice the system MTBF. The average availability over a mission also consists of the same constant term and a declining term. The second term of equation (2.18) does not fall quite as rapidly as that of equation (2.16) but still becomes negligible at about three or four times the MTBF.

Estimates based on the steady-state availability always err on the side of pessimism but are seldom subject to errors in excess of the accuracy of the basic data. First missions might be thought an exception, but in practice any increase due to newness tends to be offset by failures due to initially defective material (quality faults). Components of complex systems, such as spacecraft and manufacturing systems, are usually run for so long on tests of various kinds that the system is far from new on its first mission. The assumption of pseudo-Poisson failures is therefore often justified. Good design practice also dictates that components with low reliability should have high maintainability. If MTBF is low/high then MTTR will also usually be low/high. If we assume a rough proportionality between individual item failure and repair times and the system failure times are exponentially distributed (as they will be in a complex system), then the system repair times will also be roughly exponentially distributed, whatever the form of the component repair time distributions (usually lognormal). Although there are a lot of 'ifs' in the reasoning above, it does show that the exponential model for both distributions at system level is likely to be good enough for practical purposes. More importantly, we have also demonstrated that assumptions about distribution forms are only necessary for time-dependent calculations.

2.3 WORKED PROBLEMS

PROBLEM 2.1

The system in Table 2.1 consists of seven items having the following MTBFs and MTTRs. Find the availability of each item and of the system and the MTBF and MTTR of the system as a whole.

ANSWER

First we calculate the item availabilities from $A = \theta/(\theta + \phi)$. These are shown for convenience in the last column of Table 2.1. The system availability is then found from equation (2.11) to be

$$A_{sys} = 0.990\,67.$$

Table 2.1 Data for Problem 1

Item i	*MTBF(h)* θ	*MTTR(h)* δ	*Availability* A_i
1	2 000	2	0.999 00
2	3 000	6	0.998 00
3	5 500	1	0.999 82
4	10 000	22	0.997 80
5	10^6	100	0.999 90
6	30	0.1	0.996 68
7	500	0.3	0.999 40

The system MTBF is found from equation (2.3) to be

$$\theta_{sys} = 27.435\,\text{h}.$$

Note how close θ_{sys} is to the extraordinarily low figure of 30 h for item 6.

The system MTTR is then found from equation (2.7) to be

$$\phi_{sys} = 0.2583\,\text{h}.$$

Note how items 6 and 7 dominate the calculation of ϕ_{sys}.

As a check we calculate $\theta_{sys}/(\theta_{sys} + \phi_{sys})$ and find that once more

$$A_{sys} = 0.990\,67.$$

PROBLEM 2.2

In a civilian airliner there are 372 separately identified items which must all be available at take-off. The geometric mean of their availability is 0.9999. Find the percentage of take-offs which will be conditional on completion of repairs to one or more items.

ANSWER

The provision of the geometric mean of their availabilities presupposes a product rule thus

$$A = 0.9999^{372} = 0.963\,48,$$

so the required percentage is 3.652%.

Another approximation would be to take

$$1 - A \approx (1 - 0.9999) \times 372 \times 100 = 3.72\%.$$

By equation (2.11) we obtain

$$A = 1/[1 + 372 \times 0.0001/0.9999] = 0.9641,$$

giving 3.576%.

Which answer is correct? Well, equation (2.11) conditions do not strictly apply; for example the engines can run when the radar is failed and probably do. Presumably all 372 items must check out as correct, in which case we are dealing with that number of independent probabilities so the correct answer is 3.652%. The 3.72% answer is wrong because it falsely assumes that all failures occur singly whereas only one takeoff is affected when two or more items fail to check out simultaneously. Another point is, what is meant by 'available' in this context?

PROBLEM 2.3

A system consists of four items in series. The first item has a Weibull failure-time distribution with shape parameter 4 and characteristic life 7000 h. Its mean repair time is 5 h. The second has a normal distribution of failure times with mean 3000 h and s.d 200 h. Its MTTR is 3 h. The third has a constant hazard rate of 0.000 01/h and an MTTR of 1.5 h. The fourth item is a switch which works 99 times out of 100 and can be renewed in 30 minutes.

1. Find the probability that the system starts when set up new and runs for 2500 h.
2. If it was definitely not new how would you modify your answer to 1?
3. Find the system MTBF, MTTR and steady-state availability.

ANSWER

1. $$R_1(t) = e^{-\left(\frac{2500}{7000}\right)^4} = 0.983\,86$$

$$R_2(t) = \Phi\left(\frac{3000 - 2500}{200}\right) = 0.993\,79 \text{ (from normal tables)}$$

$$R_3(t) = e^{-2500 \times 0.000\,01} = 0.975\,31.$$

The product of these is $R_{1,2,3}(t) = 0.953\,61$. The probability the switch works once is 0.99 so $R_{sys}(t) = 0.953\,61 \times 0.99 = 0.944\,07$.

2. If the system was definitely not new, we would not know what points in their distributions had been reached by each item. The best we can do is assume that the system as a whole behaves to a Poisson law with MTBF given by

$$\theta_{sys} = 1 \Big/ \sum_{1}^{3} (1/\theta_i)$$

in which

$$\theta_1 = \eta\Gamma(1 + 1/\beta) = 7000 \times \Gamma 1.25 = 7000 \times 0.906\,40 = 63\,448\,\text{h}$$
$$\theta_2 = 3000\,\text{h}$$
$$\theta_3 = 1/0.000\,01 = 100\,000\,\text{h}.$$

For the first three items $\theta_{1,2,3}$ is 2785 h. Then $R_{1,2,3}\ (\Delta T) \approx e^{-2500/2785}$ = 0.4075. Then multiply by 0.99 for the switch, giving

$$R_{sys}(\Delta T) = 0.4034.$$

The considerable difference arises because we do not know what the ages of items 1 and 2 are at the start of the period of duration ΔT = 2500 h so we have to assume that it is overall a Poisson process. With only a small number of items in the system, the curve of ROCOF will be very wavy but will have a mean value of $1/\theta_{sys}$ as calculated above. The best guess we can make is one that will be correct on average, which is what the calculation above does. It is also worth noting how far in error the answer to 1 would be if the only data available were the item MTBFs or failure rates. In preparing a system like this for an important mission, substantial gains in expected mission reliability can be achieved by preventively renewing key components, provided of course that this maintenance work is carried out without error, and/or is thoroughly checked out before the mission starts.

3. θ_{sys} = 2785 h as calculated above, but we must also make allowance for the switch reliability. As we do not know the mean time between switch operations we cannot do this. However, it is unlikely to affect the value of θ_{sys} very much unless the missions are very short. If the mean mission length is 2500 h as in part 2 then the equivalent MTBF for the switch is 2500/(1−99/100) = 250 000 h. This reduces θ_{sys} to 2754 h.

   We can calculate ϕ_{sys} from equation (2.7)

$$\phi_{sys} = 2754\,[5/6344 \times 8 + 3/3000 + 1.5 \times 0.00001 + 1/2 \times 250\,000]$$
$$= 4.971\,\text{h}.$$

Combining these answers

$$A_{sys} = 2754/(2754 + 4.971)$$
$$= 0.9978.$$

This represents nearly 19 hours downtime in each year of attempted full-time operation. It is often salutary to make such calculations because the availability figures look so much better when expressed as decimals or percentages than the same information presented as hours lost per year (which any executive can easily translate into lost money!)

CHAPTER 3

Active parallel redundancy

3.1 ACTIVE AND STANDBY DISTINGUISHED

Redundancy was defined in Chapter 1 as the provision of more items than are actually required for satisfactory operation for the purpose of improving reliability. There are two principal forms, active (or active parallel) and stand-by. In active redundancy, as the name implies, all the items that are for the time being available are operated, but the system can continue to provide the required service with as few as m out of n of them operating. The items are then said to be in m-out-of-n active parallel redundancy. In stand-by redundancy, any items beyond the required number m are shut down whether failed or not. In both cases the redundant sub-system is regarded as failed only when less than m items remain for the moment in working order. Redundancy is often used to improve system reliability or availability in the absence of more reliable single items. Thus, a busy office manager might decide to have two photocopiers, not because the load is too great for only one machine even at peak times but simply because no manufacturer seems to make machines that fail less than three times a month, and no service operation will promise to come and fix it the same day in time for the last postal collection.

Frequently, in a predominantly series system there is one item which is significantly less reliable than the rest. Such items are the first candidates for redundancy. The improvement is dramatic, but of course the cost is high.

3.2 RELIABILITY OF ACTIVE PARALLEL SYSTEMS

3.2.1 One-shot devices

The distinction between active and standby becomes a little unclear in the case of one-shot devices, but can best be illustrated with reference to a hunter shooting a rabbit with a double-barrelled shot-gun. If he fires both barrels simultaneously, apart from the danger of injury to the hunter's shoulder, he is practising active parallel redundancy in his use of the gun. If he fires one shot and waits for the result, and only fires the second barrel if the rabbit fails to fall over then he is using the barrels in standby.

The probability that the hunter misses with both barrels when he fires in parallel is given by

$$P_2 = p^2, \tag{3.1}$$

where p is the probability that he misses with one barrel. Turning the equation round so that it is in terms probability of success or *reliability*

$$R_2 = 1 - (1 - R)^2. \tag{3.2}$$

Generalizing for a one-out-of-n parallel system of dissimilar items

$$R_{\text{sys}} = 1 - \prod_{i=1}^{n}(1 - R_i). \tag{3.3}$$

Most parallel systems are made up of identical items, in which case the number that operate successfully on demand is distributed binomially. If at least m must operate for system success then the system reliability is given by

$$R_{\text{sys}} = \sum_{i=0}^{n-m}\left[\binom{n}{i} \cdot R^{n-i} \cdot (1 - R)^i\right], \tag{3.4}$$

where $\binom{n}{i}$ is the combinatorial, $n!/i!(n - i)!$ This is the probability that *m or more* out of n operate when tried. The individual terms are the probabilities of *exactly* $n - i$ items operating.

3.2.2 Time-dependent reliability

If the items are non-identical, then we may substitute $R_i(t)$ for R_i in (3.3) above. For identical items, the same substitution can be made in (3.3) or (3.4) as applicable. The general equations are therefore for one-out-of-n

$$R_{\text{sys}}(t) = 1 - \prod_{i=1}^{n}[1 - R_i(t)], \tag{3.5}$$

and for m-out-of-n identical

$$R_{\text{sys}}(t) = \sum_{i=0}^{n-m}\left[\binom{n}{i}R(t)^{n-i}\{1 - R(t)\}^i\right]. \tag{3.6}$$

Note that once again neither equation assumes any particular distribution form; they are both valid for any distribution or none. If all the distributions are exponential, then the equation may refer to any interval T_1 to T_2 which starts with all items running, i.e. substitute $T_2 - T_1$ for t in (3.5) for $m = 1$ or in (3.6) for $m > 1$ if items are all identical. For other distributions the items must be new or else their ages at the start of the 'mission' period t must be known. In this context, identical items of different ages must be treated as different items.

Table 3.1 Probability table for two-out-of-four system

Event	*Probability*
All four items succeed	$R_1R_2R_3R_4$
Item 1 fails; 2,3,4 succeed	$(1 - R_1)R_2R_3R_4$
Item 2 fails; 1,3,4 succeed	$R_1(1 - R_2)R_3R_4$
Item 3 fails; 1,2,4 succeed	$R_1R_2(1 - R_3)R_4$
Item 4 fails; 1,2,3 succeed	$R_1R_2R_3(1 - R_4)$
1,2 fail; 3,4 succeed	$(1 - R_1)(1 - R_2)R_3R_4$
1,3 fail; 2,4 succeed	$(1 - \mathrm{R}_1)R_2(1 - R_3)R_4$
etc.	

If the items are different and m exceeds 1, calculation becomes less simple. The probability of every possible combination of failed and successful items at the end of time t has to be considered. The system reliability is then the sum of the probabilities of the successful end-states. Consider, for example, a two-out-of-four parallel system (Table 3.1) in which all the items are expected to behave differently, either because they are of different designs or because they start the mission with different ages.

The R_i above may be one-shot probabilities, time-from-new reliabilities or conditional interval reliabilities. If $m \geqslant n/2$ then it is easier to calculate the system unreliability and subtract from unity. This is so in the example above where there are only four ways that three items can be failed plus one for all four. This reduces the number of computations. If all the probabilities are calculated then there is a check on accuracy (they should all add to unity), as well as more complete information.

3.2.3 Variation of MTTF with load

It is usually assumed that the failure time distributions of individual items are not affected by the presence of others in parallel. In fact, of course, load-sharing may well reduce the failure propensity of all the items in parallel. In mechanical systems, stresses vary as the square of the speed of operation and increased stress can have very dramatic effects upon MTTF. In electronic systems, m may effectively be determined by sudden failure of all remaining items due to excessive current. A frequent assumption, of no particular merit relative to the physics of failure, is that MTTF varies inversely with load. This is a poor approximation: for example, if the total current is shared equally among remaining electronic parallel items. In this case the heat to be dissipated by each component increases as the square of the current it has to carry and the hazard rate increases exponentially with absolute temperature (Arhenius law, MIL HDBK 217). The law followed by the MTTF relative to the number remaining will depend upon the cooling arrangements but is unlikely to be as simple as is usually assumed. Whatever the form of variation assumed, it becomes impossible to develop expressions for stage reliability which do not involve the distribution form.

Consider a two-out-of-three parallel system in which the items are identical and have an exponential failure time distribution. Let θ be the item MTTF when all three items are operating. This falls to $2\theta/3$ when one item has failed on the assumption postulated above. Because the MTTF is reduced for the remainder of the interval when a failure occurs and we have no way of knowing when this may happen within the interval T it is *not* correct to write

$$R_{sys}(T) = \exp(-3T/\theta) + 3\exp(-2 \times 3T/2\theta)[1 - \exp(-T/\theta)].$$

Using methods developed in Chapter 5, the reliability for exponential distribution, works out to be

$$R_{sys}(T) = (1 + 3T/\theta)\exp(-3T/\theta) \tag{3.7}$$

(which just happens to be the same as that of a three-out-of-four standby system with individual MTTF θ).

If the MTTF were unaffected, then the exponent in the second term would be $-2T/\theta$ instead of $-3T/\theta$. The difference can be considerable even for this rather cautious assumption, as a few trial calculations will quickly demonstrate. In practice, designers use the MTTF for the most heavily loaded condition so that calculations based upon simple binomial logic are pessimistic rather than optimistic about the system reliability. Alternatively, the field data upon which θ is calculated include failures under all the possible stress conditions and can therefore be taken as a properly weighted mean value such that the simple binomial logic is applicable to an overall calculation, although not to a conditional situation in which it was known that one item had already failed.

3.3 AVAILABILITY OF ACTIVE PARALLEL SYSTEMS

3.3.1 Introduction

In this section it will be assumed that only steady-state answers are required. Except where stated otherwise, it is also assumed that all repairs start immediately following failure and do not prevent operation of the remaining items. Other situations can be dealt with, but not without making assumptions about the forms of distribution of failure and repair times. These other situations are left to a later chapter. In availability calculations, it is also necessary to be clear about whether the system is stopped when less than m out of n items are available, or continues to operate with whatever items remain. Productive plant usually continues at reduced output, indeed the concept of good or failed being the only conditions of the system (which is implicit in simple theory) may be inappropriate. In any case, failure is what we define it to be, and what we are to calculate here is the probability that the system is not in a defined failed state at a random instant in its life.

For example, a chemical plant might have to be stopped below a certain level because low throughput was dangerous. But an airline pilot faced with triple engine failure would not shut down the fourth engine even though the aircraft was incapable of sustained straight and level flight on only one engine. Any power is better than none in his situation, and at least the lights would stay on to the end. Nervous passengers may care to note that in fact most aircraft *will* fly on one engine but only if already fully airborne. The principal dangers in air travel arise from overloading, at takeoff and landing, and latterly from bombs. When more than one engine does fail in flight it is usually due to a common cause which defeats redundancy, such as ingestion of a flock of birds or volcanic ash particles. A common cause can be treated for calculation purposes as an additional item in series with the parallel arrangement. A formerly serious common cause was control failure which has been largely overcome by redundancy, diversity and separate routing of control lines. You are nowadays safer in an airliner than a motor car.

3.3.2 *m*-out-of-*n*, continue on system failure

In this case, the system runs on as long as there is at least one item available, but is regarded as failed if less than m are available. All the states suggested by the binomial expansion of scale n can exist and so the availability expression is very similar to that for reliability, i.e.

$$A_{\text{sys}} = \sum_{i=0}^{n-m} \binom{n}{i} a^{n-i}(1-a)^i, \tag{3.8}$$

where a is the item steady-state availability.

3.3.3 *m*-out-of-*n*, stop on system failure

In this case, the system is stopped as soon as the pre-defined failed condition is reached, i.e. whenever there are $n - m + 1$ failures. In contrast to the situation leading to equation (3.8) above, there is now only one downstate of the system. The upstates and this one downstate now occupy all the system time between them, so their combined probability must be unity. The probabilities of system states suggested by the binomial expansion for the states which still exist under the changed operating conditions are still in the correct *ratios* relative to one another, but their *sum* is too small. This is adjusted by dividing each of them by one minus the sum of the erstwhile probabilities of the non-existent states

$$A_{sys} = \sum_{i=0}^{n-m} \binom{n}{i} a^{n-i}(1-a)^i \Big/ \left[1 - \sum_{j=0}^{m-2} \binom{n}{j} a^j (1-a)^{n-j}\right]. \tag{3.9}$$

Note that the system has $n - m + 1$ up states plus one downstate and that $m - 1$ states of the binomial expansion do not exist.

3.4 MEAN AND VARIANCE OF TIMES TO FAILURE OF PARALLEL SYSTEMS

3.4.1 MTTF of systems of non-repairable items

The system may exist in a number of states characterized by the number and identity of items still working. In each of these states there is a unique distribution of times from the failure of one item to the next failure of an item. The probability-sum of the means (variances) of these distributions is the mean (variance) of the stage or system failure time distribution. If the items are different then there are a great many such states, one for each number of items surviving and each order of failing. For more than, say, three items it becomes too tedious to calculate all the probabilities, means and variances so that the overall mean and variance can be found. However, in most cases the items will be identical, which makes calculations easier because the number of effectively different states is reduced. If all the item hazard rates or ROCOFs are constant, then calculation becomes even easier. Consider a one-out-of-two active parallel system without repair in which the items A and B are different. From first principles, the combined p.d.f. is the sum of the individual p.d.f.s each weighted by the probability that the other item fails first, i.e.

$$f_{A+B}(t) = f_A(t)F_B(t) + f_B(t)F_A(t). \tag{3.10a}$$

The mean time to first system failure is then the first moment of this combined distribution, i.e.

$$\theta_{sys} = \int_0^\infty t f_{A+B}(t)\mathrm{d}t. \tag{3.10b}$$

This equation has no distribution-free solution.

An approximation, which is exact in the case of exponential distributions, is obtained most easily by matrix methods described in Chapter 5, but can also be derived from (3.10b)

$$\begin{aligned}\theta_{sys} &= \theta_A + \theta_B - \theta_A\theta_B/(\theta_A + \theta_B)\\ &= 1/\lambda_A + 1/\lambda_B - 1/(\lambda_A + \lambda_B),\end{aligned} \tag{3.10c}$$

where $\lambda = 1/\theta$. In words, θ_{sys} is the sum of the MTFFs of both (all) the parallel items, less the mean time during which both (more than one) items run. The bracketed observations apply to the one-out-of-n case. Variances can be calculated in the same way for the exponential case.

3.4.2 Repairable system MTBF and MTTR

A repairable system, whether active parallel or not, experiences a series of cycles of varying length, each consisting of a period of uptime followed by a period of downtime. The ratio of the means of these periods is given by

$$\rho_{sys} = \theta_{sys}/\phi_{sys} = A_{sys}/(1 - A_{sys}), \tag{3.11}$$

where A_{sys} is the system availability. Now, in any active parallel system of identical items, A_{sys} can be calculated by the distribution-free methods given above in equations (3.8), *et seq*. By rearranging (3.11), the system MTBF or MTTR can be calculated if the other is known. In most cases the MTBF of individual items is much greater than the MTTR and little error will be introduced by taking the repair times as exponentially distributed. If this is done it is often easier to estimate the system MTTR. If one repair will put the system back in operation then the system MTTR in the exponential case is simply the item MTTR divided by the number of simultaneous repair operations in progress in the single-system downstate. Thus, this method is easy to operate if the system is stopped on failure, but not if it continues or has, for any other reason, more than one downstate. Without the exponential assumption the mean rate of completion of repairs is not constant but varies with the times since each of these repairs began. These times in turn will depend upon both the repair-time and the failure-time distributions.

For example, consider a two-out-of-four active parallel system which stops on system failure and has three repairmen. The first point to note is that there will never be more than three repairs required simultaneously, so the binomial ratios of state probabilities will not be upset by lack of repair staff. The system availability is

$$A_{sys} = [a^4 + 4a^3(1 - a) + 6a^2(1 - a)^2]/[1 - (1 - a)^4]. \tag{3.12}$$

The system MTTR under the exponential assumption is $\phi/3$. Then the system MTBF is

$$\theta_{sys} = \phi_{sys}A_{sys}/(1 - A_{sys})$$

$$\theta_{sys} = \phi[a^4 + 4a^3(1 - a) + 6a^2(1 - a)^2]/ 3[4a(1 - a)^3 + (1 - a)^4]. \tag{3.13}$$

It has not to the authors' knowledge previously been noted that a reasonably accurate estimate of the system MTBF can be obtained simply by assuming that the repair times are exponentially distributed. The form of the failure-time distribution is unimportant and results, previously thought to depend upon constant rates for both failures and repairs, have been shown to be widely useable whatever the distribution forms.

Note that had there only been two repairment the state probability ratios would not have been strictly binomial and so the substitution of $\phi/2$ for $\phi/3$ would not have given a correct answer. Nor would (3.13) give a correct answer if all repairs were left until system failure. These problems can be solved using matrix algebra and Markov theory. This more advanced work is left to Chapters 5 and 6. However, there are a few other special cases which do not depend upon the assumption made in Chapter 5, namely that failure and repair distributions are all exponential in form so that rates of transition between system states are independent of the time spent in the current state. If all repairs are left until system failure then all must be repaired before starting the system again. Otherwise there is a permanent loss of redundancy. In the case where $m = 1$ and only one repair can be progressed at a time ($r = 1$), there is actually no gain in availability at all, both r and ϕ are multiplied by n to obtain the system figures. In an m-out-of-n system with no repairs until system failure and all items new at the start of the assessment period, the system MTTFF (mean time to first failure) can be found by inverting the mean rank (see Chapter 12 for explanation of this term) of the $n - m + 1$th failure out of n. That is, the last item failure which constitutes the system failure occurs on average at a point on the item failure-time distribution corresponding to a known probability which is the same for any continuous distribution, thus

$$F(\theta_{sys}) = (n - m + 1)/(n + 1). \tag{3.14}$$

If the system is mature and no repairs occur until system failure in an m-out-of-n parallel system it is best to assume that all the individual failure distributions are exponential. In this case the time to the first item failure averages θ/n and the next $\theta/(n - 1)$, and so on, giving

$$\theta_{sys} = \sum_{i=m}^{n} \theta \Big/ i. \tag{3.15}$$

If only one repair is possible at a time then

$$\phi_{sys} = (n - m + 1)\phi. \tag{3.16}$$

If there are r repair teams, $r \leqslant n - m + 1$, then the mean time to the first repair is ϕ/r. The mean time to the next will also be ϕ/r unless there are less than r still to be repaired in which case it will be $\phi/(r - 1)$ and the next $\phi/(r - 2)$, etc. This logic gives

$$\phi_{sys} = (n - m + 2 - \mathrm{r})\phi/r + \sum_{i=1}^{r-1} \phi \Big/ (r - i). \tag{3.17}$$

Note that equation (3.17) assumes exponential repair times. When no repairs take place until system failure the system is often rather confusingly described as 'unmaintained', whereas one in which repairs occur whilst the system is running is called 'maintained'.

3.5 WORKED PROBLEMS

PROBLEM 3.1

An airliner has four engines all of which are needed for takeoff and at least two for a safe landing. Failures from internal causes occur on average every 2500 h despite preventive maintenance. If the average flying time is 10 h and each takeoff is equivalent to 6 h cruising, find the proportion of landings attempted on one engine only.

ANSWER

As the engines are complex maintained items it will be reasonable to assume Poisson failures, i.e.

$$R(t) = \exp(-t/2500).$$

In this case the effective t is 16 h for each flight

$$R(t) = 0.993\,62.$$

The probability that three out of four fail during a flight

$$\begin{aligned} &= 4R(t)[1 - R(t)]^3 \\ &= 1.0 \times 10^{-6}. \end{aligned}$$

(In addition there will be $[1 - R(t)]^4$ chance of complete loss of all power, this is about 2.6×1.0^{-9}.)

COMMENTS

1. Airliners are more likely to encounter common cause failures such as bird-strikes than multiple loss of engines by pure binomially calculated chance.
2. A proportion of the multiple failures will occur at takeoff (6/16). In such cases two failures can be fatal.
3. Actual risks are increased by common-mode failures but are decreased because failed engines may restart temporarily to effect a safe landing.
4. Airlines actually achieve better statistics than these.

Bearing in mind comment 2 we now calculate the probability of double failure due to random as opposed to common causes during takeoff, that is

$$6R(t)^2[1 - R(t)]^2,$$

but t is now effectively 6 hours. This works out to

$$2.411 \times 10^{-4}$$

or about 1 takeoff in 4000. Probably at least half of these will occur in time for the pilot to abort the takeoff safely, but even so the risk is an order greater than at landing.

PROBLEM 3.2

A high-speed passenger train has an engine at each end. Its speed is reduced from 200 km/h to 150 km/h if one engine should fail. If the journey is scheduled at 180 km/h over a total distance of 600 km, how late will the train be on average if engines fail on average every 2000 h despite regular maintenance?

ANSWER

The maximum speed is irrelevant to the calculation. Both engines will remain operating on most journeys and we can find the proportion as $R(t)^2$, where $t = 600/180 = 3\frac{1}{3}$ h and

$$R(t) = \exp(-t/2000) = 0.998\,33$$
$$R(t)^2 = 0.996\,67.$$

If a failure does occur with probability

$$2R(t)[1 - R(t)] = 0.003\,325,$$

then on average it will take place halfway. From halfway to journey's end at 180 km/h would take 1 h 40 min but at 150 km/h it would take 300/150 = 2 h so the train would be 20 minutes late on average in 0.3325% of cases. The average lateness is therefore

$$20 \times 0.003\,25 = 0.0665\,\text{min}.$$

COMMENTS

1. This answer is not helpful to travellers who would be more interested to know that even on one engine all the way it would not be more than 40 minutes late.
2. The calculation would perhaps serve to show whether or not engine problems were a major cause of lateness in trains when this figure is compared with the average lateness actually achieved.
3. The proportion of journeys not completed or very late due to double failure is 2.8×10^{-6}.
4. In Problem 3.1 we were more worried about the safety of the aircraft, while safety tends to be taken for granted in trains where the main concern is punctuality.

PROBLEM 3.3

At a certain range and specified target a single missile fired at an attacking aircraft has a kill probability of 0.86. How many missiles must be fired if there is not time to fire again after observing the effects of the first salvo and we wish to be 99% confident of a kill?

ANSWER

Let the number required be n. Then

$$(1 - 0.86)^n = 1 - 0.99$$
$$n \log 0.14 = \log 0.01$$
$$n = 2.34.$$

Since we cannot fire 0.34 of a missile we actually need to fire three. Compare

$$\text{two missiles } R = 1 - (1 - 0.86)^2 = 0.9804$$
$$\text{three missiles } R = 1 - (1 - 0.86)^3 = 0.9973,$$

so about 997 attacks in 1000 will be successfully repelled if we fire three missiles.

PROBLEM 3.4

Find the reliability to 1000 h, the steady-state availability and the system MTBF and MTTR for a three-out-of-five active parallel system consisting of identical items which fail according to a Weibull distribution with parameters $\beta = 2.6$, $\eta = 3000$ h and having an MTTR = 7 h.

ANSWER

The individual item reliability function is

$$R(t) = \exp[-(t/3000)^{2.6}],$$

which for $t = 1000$ gives

$$R(1000) = 0.944\,15.$$

The reliability to 1000 h of the system is

$$\begin{aligned} R_{\text{sys}}(1000) &= \sum_{i=0}^{5-3} \binom{5}{i} R(1000)^{5-i}[1 - R(10000)]^i \\ &= R(1000)^5 + 5R(1000)^4[1 - R(1000)] \\ &\quad + 10R(1000)^3[1 - R(1000)]^2 \\ &= 0.9984. \end{aligned}$$

The individual MTBF is

$$\theta = \eta\Gamma(1 + 1/\beta) = 3000 \times \Gamma \times 1.3846 = 2666\,\text{h}.$$

Assuming three repair crews and exponentially distributed repair times so that no repair has to await attention, clearly

$$\phi_{\text{sys}} = \phi/3 = 2\tfrac{1}{3}\,\text{h}.$$

The availability of a single item is

$$a = 2666/2673 - 0.997\,38$$

$$A_{sys} = [a^5 + 5a^4(1 - \text{a}) + 10a^3(1 - a)^2]/[1 - 5a(1 - a)^4 - (1 - a)^5]$$
$$= 0.999\,999\,8$$

$$\theta_{sys} = \phi_{sys}A_{sys}/(1 - A_{sys}) = 11.67 \times 10^6\,\text{h}.$$

It has been assumed that repairs take place as required in arriving at the answers above. If the system is repaired only after three items have failed and by a single repair team, and if all repairs are (statistically equivalent to) renewals, the answers are very different. The third failure out of five will occur on average at a point on the individual item failure time distribution where with $n = 5$, $m = 3$,

$$F(t) = 1 - R(t) = (n - m + 1)/(n + 1) = 0.5,$$

so to obtain θ_{sys} we have to solve

$$\exp[-(\theta_{sys}/3000)^{2.6}] = 0.5$$
$$-(\theta_{sys}/3000)^{2.6} = \log 0.5$$
$$2.6 \log(\theta_{sys}/3000) = \log(-\log 0.5)$$
$$\theta_{sys} = 2605\,\text{h}.$$

There would be three to repair so ϕ_{sys} would be $3 \times 7 = 21\,\text{h}$, giving an availability of

$$A_{sys} = 2605/2626 = 0.9920.$$

This example illustrates the widely differing results which arise from changes to operating and maintenance policies for the same basic system.

It is also instructive to compare the MTTFF found above with that for an exponential distribution with the same item mean. This would give

$$\theta_{sys} = \theta(1/5 + 1/4 + 1/3)$$
$$= 2666 \times 0.7833 = 2088\,\text{h}\ (c.f.\ 2605)$$

which is a considerable underestimate leading also to an inaccurate estimate of availability of

$$A_{sys} = 2088/2109 = 0.9902\ (c.f.\ 0.9920)$$

implying 16 h more downtime per year than would be realized on average.

REFERENCE

MIL-HDBK 217 (regularly updated), *Reliability Prediction of Electronic Equipment*, RomAir Development Center, Griffis Air Force Base, NY 13441.

CHAPTER 4

Standby redundancy

Other things being equal, and given reliable switching, standby redundancy has obvious advantages over active parallel redundancy. Except in the case of one-shot devices, where the reliabilities are equal, it also gives higher reliabilities and availabilities. However, other things are seldom equal. Parallel redundancy is preferred when switching reliability is sufficiently low relative to item reliability to negate the natural advantage of standby, and also when no interruption of service can be tolerated even for an instant whilst the changeover is made. This results in parallel redundancy being used in electronic systems and standby mainly in mechanical systems. There is, of course, also the possibility that the standby will not start when required or is only discovered to be failed when one tries to bring it into service. A partial solution to this last problem is the so-called 'warm standby' which is not completely shut down but 'idling' until needed. In this intermediate state it has a finite MTTF or MTBF greater than that when on load, whereas a true standby, *by definition*, cannot fail whilst not in use.

It is unusual for m-out-of-n standby systems to exceed $m = 2$ or $n = 3$ but the authors know of several systems with m up to 7 and n up to 8. Standbys are often deliberately different from the preferred item, either for safety, particularly in nuclear installations, or else because the old machine has been retained as a standby for the new, typically in manufacturing systems.

4.1 RELIABILITY OF STANDBY REDUNDANT SYSTEMS OF ONE-SHOT DEVICES

4.1.1 One-shot devices – one-out-of-n

Consider two devices, A and B, either of which can perform a certain function. Let the probabilities of correct function be R_A and R_B and the overall probability of a correct result be R_{sys}. Let A be preferred to B. Then the system operates correctly if either A functions, or A having failed B is tried and functions correctly. Thus

$$
\begin{aligned}
R_{sys} &= R_A + (1 - R_A)R_B \\
&= R_B + (1 - R_B)R_A \\
&= 1 - (1 - R_A)(1 - R_D).
\end{aligned} \tag{4.1}
$$

From this analysis we may conclude that the order of trying the items does not affect the system reliability and that the result is the same as for an active parallel system. Although there is no gain in reliability over a parallel system, the average number of items used per successful attempt is reduced if standby operation is employed. This average will be minimized if the most reliable item is tried first, with the others following in descending order of reliability.

For n items, one only required for success and tried in the order $1, 2, \ldots, n-1, n$

$$
\begin{aligned}
R_{sys} &= R_1 + (1 - R_1)R_2 + (1 - R_1)(1 - R_2)R_3 \ldots \\
&= R_1\left[1 + \sum_{i=2}^{n}\left\{R_i\prod_{j=1}^{i-1}(1 - R_j)\right\}\right] \\
&= 1 - \prod_{i=1}^{n}(1 - R_i).
\end{aligned} \tag{4.2}
$$

This last result is not immediately apparent, but a few trial calculations will convince doubters. It is easier to demonstrate by considering the unreliability. The system fails with probability

$$
\begin{aligned}
1 - R_{sys} = 1 - R_1 - R_2(1 - R_1) - R_3(1 - R_1)(1 - R_2) \ldots \\
- R_i\prod_{j=1}^{i-1}(1 - R_j) \ldots - R_n\prod_{j=1}^{n-1}(1 - R_j).
\end{aligned} \tag{4.2a}
$$

For one-shot devices there is no reliability advantage in standby operation over parallel, but standby operation does result in fewer items being used, which is often a more important advantage than higher reliability. The mean number used is given by

$$
\bar{X} = R_1 + 2R_2(1 - R_1) + \ldots + nR_n\prod_{j=1}^{n-1}(1 - R_j)
$$

$$
\bar{X} = R_1 + \sum_{i=2}^{n}\left[iR_i\prod_{j=1}^{i-1}(1 - R_j)\right]. \tag{4.2b}
$$

4.1.2 Identical one-shot devices – *m*-out-of-*n* standby

In this case the items are tried until either m have worked correctly or all n have been tried. It is immaterial to the probability calculation whether the first m items are tried together or in sequence.

The probability that the first m items tried all operate correctly is R^m.

There are m ways that one of the first m could fail so the next term is $mR^m(1 - R)$. There are $\binom{m+1}{2}$ ways in which two of the first $m + 1$ items tried could fail, and so on. The general, jth, term is $\binom{m+j-1}{j}R^m(1 - R)^j$. The general equation is therefore

$$R_{\text{sys}} = R^m \sum_{j=0}^{n-m} \binom{m+j-1}{j}(1 - R)^j. \tag{4.3}$$

Although it is not immediately apparent, equation (4.3) is also equivalent to the corresponding active parallel calculation, *c.f.* equation (4.2) above. For example take $m = 2$, $n = 5$ and $R = 0.95$ and it will be found that the result is 0.999 97 in both cases. Once again, the advantage lies in the use, on average, of fewer items in standby than in parallel operation. Returning to the hunter and his shotgun, if he fires the second barrel only if needed then he will certainly use less ammunition, and if all shots are equally likely to hit he will also bag the same number of rabbits over a long period. But for the same number of shells he can obtain more rabbits by adopting a standby policy than he can by firing both barrels together. Even if the second shot, at a fleeing rabbit, is less likely to hit, this remains true.

The mean number of items used in attempts to obtain m successes from a maximum of n items per attempt is

$$\bar{X} = \sum_{x=m}^{n} \left[x\binom{x-1}{x-m}R^m(1 - R)^{x-m} \right] + n \sum_{i=n-m+1}^{n} \binom{n}{i}R^{n-i}(1 - R)^i. \tag{4.4}$$

Note that at some attempts fewer than m successes will be obtained from n trials. In equation (4.4) the first term refers to successful attempts and the second to unsuccessful attempts.

4.2 RELIABILITY OF STANDBY REDUNDANT SYSTEMS WITH TIME INVOLVED

4.2.1 One-out-of-two standby

If the system operates continuously and the failure probabilities of the items are functions of time from new or previous repair, there is a more complicated problem to be solved. Suppose that the reliability required is for a system of two new items, A and B, over a period of time t. We assume for the sake of argument that item A is used first and that B is the standby. We cannot solve this problem using methods already discussed because the time for which the standby item may be required to operate is a variable and will be zero with probability equal to the reliability of A over the time interval t, $R_A(t)$. A general solution can be obtained but it involves integrals which, depending on the form of the failure time distributions, may well be intractable. However, answers can always be obtained using a computer and numerical integration through Simpson's rule or similar means.

Let the first item, A, fail at time x, where x is not necessarily less than the required mission time t. The system reliability has two components, the probability that A lasts for the entire mission and the conditional probability that B survives given that A fails. Let the failure time p.d.f.s be $f_A(t)$ and $f_B(t)$, and the corresponding reliability functions be $R_A(t)$, $R_B(t)$. Then

$$\begin{aligned} R_{sys}(t) &= R_A(t) + \int_0^t \left[f_A(x) \int_{t-x}^{\infty} f_B(u)\mathrm{d}u \right] \mathrm{d}x \\ &= R_A(t) + \int_0^t [f_A(x) R_B(t-x)]\mathrm{d}x. \end{aligned} \tag{4.5}$$

The first term of equation (4.5) is the probability that the first item lasts for the whole mission. The second term is the conditional probability that the second item survives the mission given that the first fails. The double integral is needed because the first item could fail at any instant between 0 and t. In general, evaluation of the second term will not be easy without a computer even if the items are identical. Luckily, standby redundancy is usually associated with maintained complex items rather than components with single failure modes. Such items tend to have constant ROCOF λ under an unchanging maintenance policy during the majority of their total lives; that is the middle portion of the bathtub curve. In such cases it is legitimate to assume a Poisson process if the items are identical. Then the probability of none or one failure in any random interval T given that both items are available at the start of said interval is Poisson with parameter $m = \lambda T$. Therefore

$$R_{sys} = (1 + \lambda T) \exp(-\lambda T). \tag{4.6}$$

Equation (4.6) can be extended to cover the case where $m = n - 1$ in an m-out-of-n standby system provided that the ROCOFs are all constant. In this case, the total ROCOF of the operating item is $m\lambda$ both before the first failure and between the first and second (final) failure thus one-out-of-n standby with constant equal hazard rates or ROCOFs. In a Poisson process, the probability of exactly n events in an interval T if the constant rate is λ is given by

$$P(n) = (\lambda T)^n \exp(-\lambda T)/n! \tag{4.7}$$

This can be used to calculate standby reliability in the case of constant ROCOF or constant hazard rate because and only because the rate of failure does not vary with time or number of prior failures. The probability that a one-out-of-n system survives a total time T without system failure or repair is the sum of the probabilities of exactly $0, 1, 2, \ldots, n-1$ failures in that time, which is

$$R_{sys}(T) = \exp(-\lambda T)\sum_{i=0}^{n-1}[(\lambda T)^i/i!] = 1 - (\lambda T)^n \exp(-\lambda T)/n! \quad (4.8)$$

For an m-out-of-n system the total ROCOF is $m\lambda$ throughout and the system is stopped after $n - m + 1$ failures so we can modify (4.8) to

$$R_{sys}(T) = \exp(-m\lambda T)\sum_{i=0}^{n-m}(m\lambda T)^i/i! \quad (4.8a)$$

Referring back to Chapter 1, equation (4.8) implies that the combined distribution for n items with constant hazard rate λ is of Erlang form; that is the overall p.d.f. is

$$f_{sys}(t) = \lambda^n t^{n-1}\exp(-\lambda t)/(n-1)! \quad (4.9)$$

Strictly speaking, the concept of p.d.f. should not be applied to processes involving ROCOF. As n increases, the system p.d.f. approaches the normal form. For $n > 3$, the normal distribution is a better approximation than the exponential distribution.

4.3 MTTF AND MTBF OF STANDBY SYSTEMS

The mean time to first system failure of a one-out-of-n standby system is clearly just the sum of the means of the individual distributions. The same applies to the MTBF of a one-out-of-n standby repairable system which is not repaired until all items have failed, *but most definitely not to systems in which individual repairs are made whilst the system continues to operate* (Chapters 5 and 6). By the central limit theorem of statistics, if the means may be added then variances are also additive, i.e.

$$\theta_{sys} = \sum_{i=1}^{n}\theta_i \quad (4.10)$$

$$\sigma_{sys}^2 = \sum_{i=1}^{n}\sigma_i^2. \quad (4.11)$$

Note that these two results are distribution free. As n becomes large the system distribution tends to the normal whatever the individual distributions.

4.3.1 *m*-out-of-*n* standby systems

For m-out-of-n systems, calculations become too complicated unless the assumption is made that the hazard rates or ROCOFs involved are constant and equal. In that case only, the mean time between individual failures as the system deteriorates towards failure is θ/m, and its variance $(\theta/m)^2$. There are $n - m + 1$ such intervals in a full system failure sequence so the combined mean and variance are

$$\theta_{sys} = \theta(n - m + 1)/m \tag{4.12}$$

$$\sigma^2_{sys} = (\theta/m)^2(n - m + 1). \tag{4.13}$$

Returning to the case where the distributions are not exponential, as items fail, their replacements in the running set of m are new. This leads to a mixture of ages in the running set. However, for large m and even larger n the ROCOF or local sum of the hazard rates tends to be constant making (4.8a), (4.12) and (4.13) more and more applicable. For small m and n, this effect is not nearly so significant. Note also that in most practical cases the items in the system will be themselves complex and repairable subsystems such as engines, pumps or generators, and will start any mission with different effective ages or even with constant ROCOF. The overall effect is to make the constant-rate results rather more widely applicable practically than might at first be thought. Readers are advised, however, to take care to ensure that a constant-rate approximation is justified in any particular case. It is also generally the case that a constant-rate approximation will lead to an underestimate of reliability rather than an overestimate if the mean is correct. Equipment which is new or recently overhauled to a standard condition by renewal of all wearing parts is likely to have a low but rising total hazard rate in respect of these new parts which could well follow a normal distribution approximately until one of them fails. This also follows from the central limit theorem.

4.4 AVAILABILITY OF STANDBY REDUNDANT SYSTEMS

The MTBF, MTTR, and Availability of systems in which individual items are repaired whilst the system continues to run require matrix calculations as described in Chapters 5 and 6. These quantities also depend upon the number of simultaneous repairs possible and on the operating policy with regard to running in nominally failed states.

Using methods described in detail in Chapters 5 and 6, it may be stated here that the availability of a one-out-of-two standby system with perfect switching and no repair restrictions is given by

$$A = 2a/(1 + a^2), \tag{4.14}$$

where $a = \theta/(\theta + \phi) = \mu/(\mu + \lambda)$ is the availability of a single unit. Strictly, equation (4.14) applies only where both failure and repair rates are constant, but in practice it may be used generally as a good approximation.

4.5 WORKED PROBLEMS

PROBLEM 4.1

A surface-to-air missile system consists of ground equipment with MTBF 2000 h and MTTR 0.5 h plus missiles which have a combined reliability and accuracy giving a kill probability of 0.91. If the first missile fired is seen to miss the target there is just time to fire another. Find the probability of killing a given target and the mean number of missiles used per target.

ANSWER

The figure needed for the ground equipment is the steady-state availability which is

$$A_g = 0.99975.$$

A_g is so high that we would not be much in error by leaving it out, but for the sake of the exercise it is included. The kill probability is then

$$P_k = A_g[R + (1 - R)R],$$

where R is 0.91. This works out to

$$P_k = 0.99165.$$

If the ground equipment is not working no missiles are fired. In 91% of the rest only one is needed and in 9% two are fired. The expected consumption per target is therefore

$$0.99975(0.91 \times 1 + 0.09 \times 2) = 1.0897 \text{ missiles per target.}$$

PROBLEM 4.2

An aircraft has two identical pumps for shifting fuel between tanks to maintain trim in flight. One pump is kept running on recirculation at all times with the other on automatic standby through a switching device which operates correctly 99 times out of 100. Find the reliability and MTBF of this system over a 10-hour flight if the running pump fails about once per 1000 h on average.

ANSWER

Apart from the switch this is a simple one-out-of-two standby with constant failure rate 0.001/h, i.e.

$$R(10) = (1 + 10 \times 0.001) \exp(-10 \times 0.001).$$

To allow for the switch probability we need to multiply the second term, which covers the operation after the standby has been actuated, by 0.99, i.e.

$$R(10) = (1 + 0.99 \times 10 \times 0.001) \exp(-0.01) = 0.99985.$$

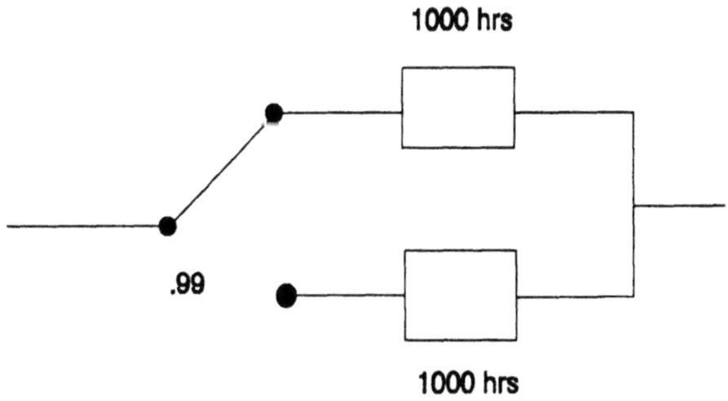

Fig. 4.1 Block diagram for the redundant fuel pump system in Problem 4.2.

The MTBF would be one average lifetime if the switch failed with probability 0.01 and two if it worked with probability 0.99, i.e.

$$1000 \times 0.01 + 2000 \times 0.99 = 1990\,\text{h}.$$

PROBLEM 4.3

A remote town is supplied with electricity by two separate power plants of different capacities and reliability characteristics as in Table 4.1.

The load varies between 4 MW max. for 12 hours at night and 12 MW max. by day. Each plant is given preventive maintenance (PM) requiring it to be off load for 12 h or 3 h as shown each week. Find the expected proportion of time for which the system is unable to supply its maximum load if the diesel is normally run only on the one night per week when the steam plant is undergoing PM.

Table 4.1 Data for Problem 4.3

	Capacity (MW)	*MTBF (h)*	*MTTR (h)*	*PM (h per week)*
Steam plant	15	3000	12	12
Diesel plant	5	2500	3	3

ANSWER

During the day, the diesel is not an effective standby (except for vital supplies) so the availability for that half of the 24 h is that of the steam plant, i.e.

$$A_s = 3000/3012 = 0.9960$$

for the diesel plant

$$A_d = 2500/2503 = 0.9988.$$

At night, the diesel is an effective standby but we have no formula except equation (4.14) to gives us the availability. However A_s and A_d are almost equal so we could with little error approximate on the basis of equation (4.14).

$$A \approx (A_s + A_d)/(1 + A_s A_d) = 0.999\,997\,6.$$

Now taking the PM requirements into consideration, for a proportion $12/(7 \times 24) = 0.071\,43$ of the time we have only the diesel and for 0.017 86 we have only the steam plant. The standby availability operates for $6/14 - 0.017\,86$ of the time and the steam plant availability for $7/14 + 0.017\,86$ assuming that the diesel is also maintained at night. The overall availability is therefore

$$\begin{aligned} A_{sys} &= A_s(0.017\,86 + 0.5) + A_d \times 0.071\,43 + 6\text{A}/14 - 0.017\,86A \\ &= 0.997\,85, \end{aligned}$$

so the system is unable to meet load for 0.215% of the time or nearly 19 hours per year. It would be better to maintain the diesel by day because it cannot meet the day load anyway. In that case

$$A_{sys} = A_s \times 0.5 + A_d \times 0.071\,43 + 6\text{A}/14 = 0.997\,92.$$

CHAPTER 5

State-space models and matrix methods

In many cases a two-state approach for reliability analysis of systems composed of multiple components is not appropriate. This is so especially when redundancy is applied or when some components have multiple failure modes having non-identical impacts on system performance. For example, consider a redundant rectifier system consisting of two diodes connected in series (Fig. 5.1(a)) and without repair. The diodes might suffer two distinct failure modes: a short-circuit failure D_2 allowing current to pass in both directions and an open failure D_3. Figure 5.1(b) shows the states and the corresponding transitions for a diode. It will be obvious that the system is only redundant to short-circuit failures and that the redundancy is not active for open failures. Even worse, the probability of an open system failure has almost been doubled. This means that the two different failure modes of the diodes have different impacts on system performance and, therefore, cannot be treated as one single failure mode. For this simple system a combinatorial solution is still possible as: $R = P_r\{Da_1Db_1 \cup Da_1Db_2 \cup Da_2Db_1\}$, however, the logic expression for system success contains dependent events since the diode states D_1 and D_2 are mutually exclusive. In Chapter 7 it is shown how to deal with systems of multistate items more generally while Chapter 8 explains how to perform probability calculations on large logical expressions with dependent events. In the rectifier example above the p.d.f.s of the two failure modes are independent and, therefore, a combinatorial solution is possible. However, combinatorial solutions are generally obstructed for those systems in which (some of) the different p.d.f.s governing component failures or repairs are dependent. For example, the lifetime of the second item of a standby redundant system starts when the first item fails and, therefore, its time of failure structurally depends on that of the first item. A similar situation occurs when a repair facility is shared by multiple items of the system; the repair of an item cannot start before the item has failed and the repair facility is free. For these cases the state-space method is more appropriate (Chapter 7). This method allows multi-state modelling for the system and its components and has a very wide applicability. The only restriction of this method is practical; the

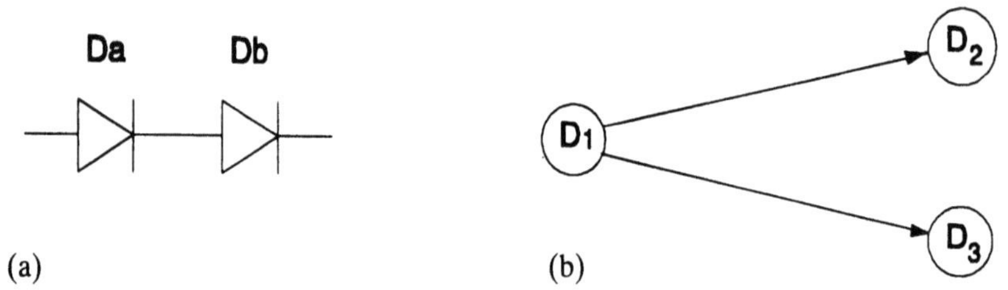

Fig. 5.1 (a) Redundant rectifier system consisting of two diodes. (b) State-transition diagram for the diodes with open and short failures.

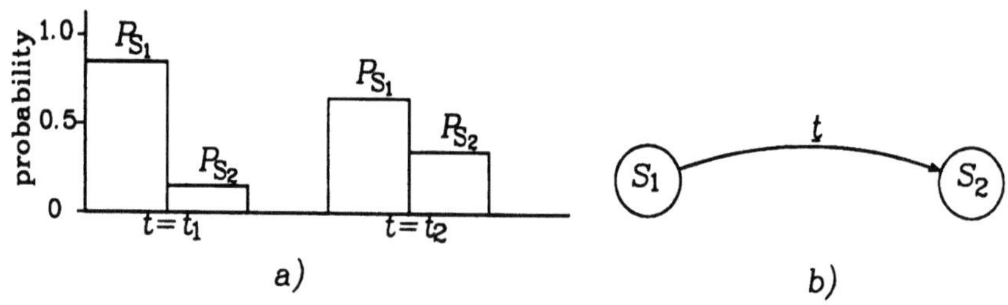

Fig. 5.2 (a) The system state S is a discrete stochastical variable. The corresponding probability distribution changes with time. (b) The state-space or Markov diagram.

number of states rapidly increases with the number of components and the number of component states involved, so that the calculations become unwieldy.

5.1 WHAT IS A STATE-SPACE MODEL?

The state-space method treats failure processes as Markovian processes which, by definition, use two stochastical variables, the state S of the system observed and the time t. Since the number of distinct system states is finite, the system state S is a discrete stochastic variable which can represent any element of the complete set of (mutually exclusive) system states. However, the time instants at which a system passes from one state to another must be described by a continuous stochastical variable which is the time t. As an example, observe a simple item such as an audio amplifier which is assumed to have two states; S_1 with acceptable output and S_2 with unacceptable or no output. The bar-diagrams in Fig. 5.2(a) show the (discrete) probability distribution of both states at different time instants. This means that the state S at any time instant is a discrete stochastical variable. The (Markov) diagram in Fig. 5.2(b) shows that the transition from state S_1 to state S_2 is governed by the instant of failure which is a continuous stochastical distribution in time.

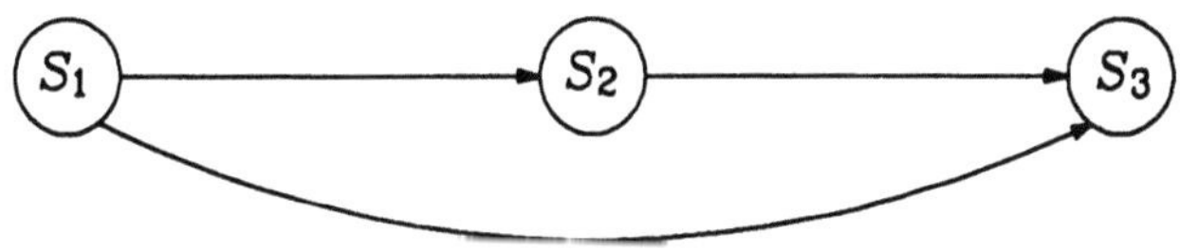

Fig. 5.3 The state-space diagram of a ball-bearing which is assumed to have three different states: S_1 – good performance; S_2 – noticeable wear-out but acceptable performance; and S_3 – failed.

It will be clear that in this simple case application of the state-space method yields no substantial advantage. However, this will be different when for practical reasons more than two system states have to be distinguished. As an example consider a ball-bearing which, from a maintenance point of view, is assumed to have three different states:

1. S_1 – good performance;
2. S_2 – increased vibration level indicating wear-out (although the performance is still acceptable replacement is necessary at the next periodic overhaul);
3. S_3 – loss of bearing function.

The state-space diagram for this component is given in Fig. 5.3 and shows the component states and the interstate transition that may occur. Starting with a new bearing (S_1) the normal wear-out process will cause a slow degeneration of the bearing resulting in a transition to state S_2 when the specified vibration level has been exceeded. When no replacements are made the degeneration process will continue until the vibration level becomes unacceptable and state S_3 has been reached. However, other failure processes resulting in a sudden catastrophic break-down of the bearing are also possible. For example when, due to a hair-line crack in one of the balls, a flake comes off smashing up the whole bearing. Therefore, a direct transition from S_1 to S_3 is added in the state-space diagram. When the stochastical distributions governing the interstate transitions are also known, the state-space model is complete.

5.2 PROCEDURE FOR STATE-SPACE MODEL CREATION

In order to set up a state-space model it is best to lay down the system states and transitions on a state-space diagram first. A simple, straightforward procedure to create state-space diagrams proceeds as follows.

1. Start at the state in which everything is perfect (S_1).
2. In this state several events may take place that affect the system's performance and/or the system configuration. These events can be of different nature such as: the failure or degeneration of components, repair or maintenance of components by a maintenance facility,

change in repair strategy, etc. For each of these events an out-going transition to a new state must be added. This new state should describe the resulting system behaviour exactly. If such a state already exists the transition can be directed towards this state and no new state has to be added, however, be sure that the states are identical, i.e. the same components are functioning (and active), identical repairs are in progress, etc. It is also possible to combine equivalent but non-identical states which arise typically when considering systems containing several identical components receiving identical treatments.

3. Repeat step 2 for all new states until it has been applied to all states of the diagram.

Although the resulting diagram completely describes the system's behaviour, a minor modification (simplification) of the diagram might be necessary if one wants to study the behaviour of a repairable system up to the first failure only, which is the case in $R(t)$, MTTFF and $z(t)$ evaluation. For these characteristics the branches leaving system-down states *must* be neglected since they describe the system behaviour after system failure and, therefore, would disturb our calculations. In this special case a further reduction of the diagram (and so of the problem) may result from combining all system-down states into one single down state.

EXAMPLE 5.1

Consider the power supply of a hospital, which, for reasons of safety, has a redundant design. When the public power supply fails, the emergency power system (EPS) takes over the supply of electricity to vital equipment. Both the public power supply and the emergency power supply may fail stochastically and independently with hazard rates $z_p(t)$ and $z_e(t)$ respectively, where t is the time elapsed since the first start of the system. It should be noted that the $z(t)$s are not identical to the (time-dependent) failure rates of the components, since after each repair to an as-good-as-new state, say t_n, the failure rate curve is repeated i.e. $z(t) = \lambda(t - t_n)$. Each supply has its own (independent) repair facility. Repair of the public power supply proceeds with a rate $z_{rp}(t) = \mu_p(t - t_{rp})$ while the repair rate of the emergency supply is $z_{re}(t) = \mu_r(t - t_{re})$, where t_{rx} is the elapsed time since the start of the repair of component x and μ_x is the repair rate function for component x. Figure 5.4 shows the step-by-step generation of the state-space diagram for the total power supply. Figure 5.4(b) describes the system up to the first system failure (no power) and can be used for evaluation of $R(t)$, $z(t)$ and MTTF (for these evaluations the states S_3 and S_4 may be combined). Figure 5.4(c) describes the system behaviour during subsequent up and down periods and can be used as a basis for the evaluation for the corresponding system characteristics such as $A(t)$, MTBF and MTTR.

Once we have created the state-space diagram and the corresponding

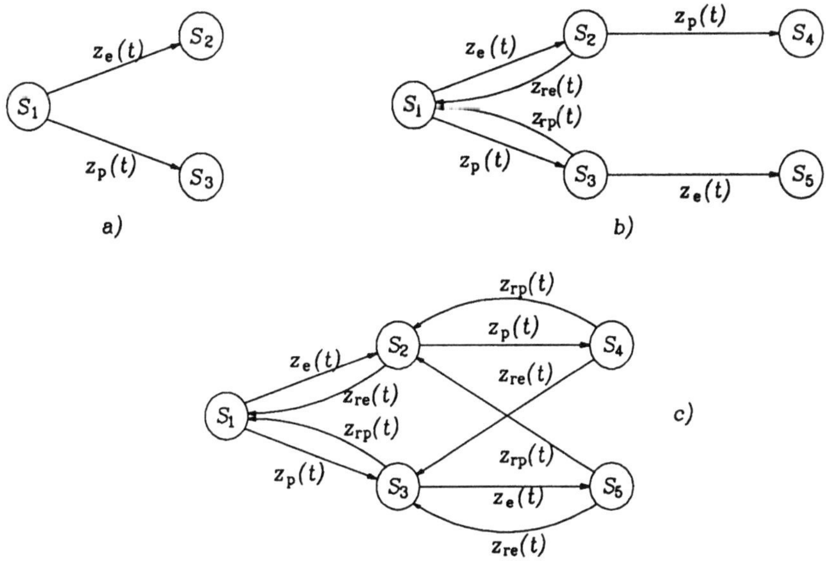

Fig. 5.4 Recursive steps of the generating process for the state-space diagram of the hospital power supply discussed in Example 5.1.

transition rates are known, a set of differential equations can be derived which constitute the mathematical representation of the state-space model. For each state of the diagram a differential equation can be derived showing that the time-derivative of the state probability equals the total frequency of entering this state through one of the incoming branches minus the total frequency of departing this state through one of the outgoing branches. In formula form this can be written as

$$\dot{P}_i(t) = \sum_{\substack{k=1 \\ k \neq i}}^{n} P_k(t) z_{ki}(t) - \sum_{\substack{k=1 \\ k \neq i}}^{n} P_i(t) z_{ik}(t) \quad (i = 1, 2, \ldots, n), \tag{5.1}$$

where $P_k(t)z_{ki}(t)$ symbolizes the unconditioned occurrence frequency of transition $S_k \rightarrow S_i$. This set of differential equations together with the initial state probabilities (at $t = 0$) constitute the system's mathematical model.

5.3 EXACT SOLUTION METHODS

Usually, the set of differential equations has an analytical solution only when all coefficients are time independent, i.e. when all transition rates are constant. In this special case the set of equations can be solved through application of the Laplace transformation rules which turn the set of differential equations into a set of linear equations. However, when one is only interested in time-independent characteristics of the system (MTTFF, MTBF, MTTR, etc.) the set of differential equations can easily be rewritten as a set of linear equations which considerably simplifies the solution method since no Laplace transformation is required. Therefore, the solution methods can be split as:

1. solution methods for time-dependent characteristics (section 5.3.1);
2. solution methods for time-independent characteristics (section 5.3.2).

5.3.1 Evaluation of transient characteristics

Evaluation of transient (time-dependent) characteristics, such as $A(t)$, $R(t)$ and $z(t)$ all require solution of the state probability functions $P_i(t)$ which can be done by means of Laplace transforms. After solving the resulting set of linear equations in the Laplace domain the state probabilities must be transformed back into the time domain again in order to obtain the actual (time-dependent) state probabilities. A more detailed description of this solution method follows.

The set of differential equations is usually written in matrix notation as $[\mathrm{d}P_i(t)/\mathrm{d}t] \equiv [\acute{P}_i] = [P_i(t)]Q_\mathrm{t}$, where Q_t is called the transition matrix. Using the basic Laplace transforms

$$\textit{Time domain}\quad \begin{array}{l}\acute{P}_i(t) \Leftrightarrow sP_i(s) - P_i(0)\\ P_i(t) \Leftrightarrow P_i(s)\end{array} \quad \textit{Laplace domain}, \tag{5.2}$$

the set of differential equations can be transformed as:

$$\left[\frac{\mathrm{d}P_i(t)}{\mathrm{d}t}\right] = [P_i(t)]Q_\mathrm{t} \Leftrightarrow [P_i(s)]sI - [P_i(0)] = [P_i(s)]Q_\mathrm{t}, \tag{5.3}$$

where I is the identity matrix and $[P_i(0)]$ is the vector of initial conditions, i.e. the vector of initial state probabilities at $t = 0$. This matrix equation can be rewritten as:

$$[P_i(s)][sI - Q_\mathrm{t}] = [P_i(s)]Q_\mathrm{s} = [P_i(0)]. \tag{5.4}$$

Introducing the inverse matrix Q_s^{-1} the equation can be rewritten as

$$[P_i(s)] = [P_i(s)]Q_\mathrm{s}Q_\mathrm{s}^{-1} = [P_i(0)]Q_\mathrm{s}^{-1}. \tag{5.5}$$

This means that solving this equation is the same as calculating the inverse matrix Q_s^{-1}. The elements of this matrix can be calculated as:

$$q_{ij}^{-1} = \frac{C_{ji}}{|Q_\mathrm{s}|} = \frac{(-1)^{j+i}M_{ji}}{|Q_\mathrm{s}|}, \tag{5.6}$$

where $|Q_\mathrm{s}|$ is the determinant of Q_s and M_{ji} is the determinant of the submatrix obtained by deleting the jth row and the ith column of the original matrix.

Usually it is assumed that the system initially starts at state S_0 which means that $P_0(0) = 1$ is the only non-zero initial state probability. Consequently, in this particular case, only the first row of the inverse matrix is of importance.

Once the state probabilities have been found, the evaluation of the $R(t)$ and $A(t)$ functions is simply the summation of the state probabilities of all system-up states, as can be seen from

$$\left.\begin{matrix} R(t) \\ A(t) \end{matrix}\right] = \sum_{i \in U} P_{s_i}(t). \tag{5.7}$$

Although the availability and reliability have identical expressions in terms of state probabilities the outcomes will be different since they result from different diagrams (section 5.2). Similarly, we can distinguish two rates of system failure: the hazard rate $z(t)$ when no repair is made after system failure and the system-down rate $\lambda_s(t)$ or ROCOF with repair after system failure. In addition to the system-down rate $\lambda_s(t)$ a system-up rate $\mu_s(t)$ can be defined. The formulae for these rates are

$$\left.\begin{aligned} \left.\begin{matrix} z(t) \\ \lambda_s(t) \end{matrix}\right] &= \frac{\sum\limits_{i \in U; j \in D} P_{s_i}(t)\lambda_{ij}}{\sum\limits_{i \in U} P_{s_i}(t)} \\ \mu_s(t) &= \frac{\sum\limits_{i \in U; j \in D} P_{s_j}(t)\lambda_{ji}}{\sum\limits_{j \in U} P_{s_j}(t)} \end{aligned}\right\} \tag{5.8}$$

EXAMPLE 5.2

When both the public power supply and the emergency power system of Example 5.1 have identical (and constant) failure rates and repair rates of respectively $\lambda = 1$/year and $\mu = 10^3$/year the diagram for reliability evaluation (Fig. 5.4(b)) can be reduced to the diagram depicted in Fig. 5.5. For each state the time derivative of its probability can be found as the sum of the frequencies of the incoming branches minus the sum of the frequencies of the outgoing branches in accordance with formula (5.1). The matrix notation for this set of differential equations then becomes

$$[\dot{P}_i(t)] = [P_i(t)] \begin{bmatrix} -2\lambda & 2\lambda & 0 \\ \mu & -(\lambda + \mu) & \lambda \\ 0 & 0 & 0 \end{bmatrix} = [P_i(t)]Q_t. \tag{5.9}$$

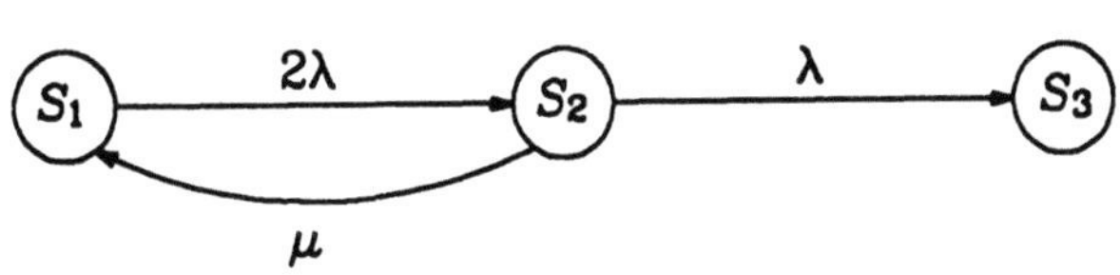

Fig. 5.5 The state-space diagram for the system in Example 5.2.

Then the Laplace domain matrix Q_s is

$$Q_s = \begin{bmatrix} s-2\lambda & 2\lambda & 0 \\ \mu & s-\lambda-\mu & \lambda \\ 0 & 0 & s \end{bmatrix}. \tag{5.10}$$

While the important submatrices M_{11} through M_{13} are found to be

$$M_{11} = \begin{bmatrix} s-\lambda-\mu & \lambda \\ 0 & s \end{bmatrix}, M_{12} = \begin{bmatrix} 2\lambda & 0 \\ 0 & s \end{bmatrix}, M_{13} = \begin{bmatrix} 2\lambda & 0 \\ s-\lambda-\mu & \lambda \end{bmatrix}. \tag{5.11}$$

Then the first row of the inverse matrix, which represents the Laplace solution of the state probabilities, can be calculated using formula (5.6) as

$$\begin{aligned} Q_s^{-1} &= \frac{1}{s^2 + s(3\lambda+\mu) + 2\lambda^2} \begin{bmatrix} s+\lambda+\mu & 2\lambda & 2\lambda^2 \\ \cdots & \cdots & \cdots \\ \cdots & \cdots & \cdots \end{bmatrix} \\ &= \frac{1}{(s+r_1)(s+r_2)} \begin{bmatrix} s+\lambda+\mu & 2\lambda & 2\lambda^2 \\ \cdots & \cdots & \cdots \\ \cdots & \cdots & \cdots \end{bmatrix} \\ \text{with} \quad r_1, r_2 &= \frac{(3\lambda+\mu) \pm \surd[(3\lambda+\mu)^2 - 8\lambda^2]}{2}. \end{aligned} \tag{5.12}$$

Prior to re-transformation into the time domain, the expressions for the state probabilities must be decomposed into partial fractions. For S_1 this results in

$$P_1(s) = q_{11}^{-1}(s) = \frac{R_1}{s+r_1} = \frac{R_2}{s+r_2}, \text{ with } \left. \begin{aligned} R_1 &= \lim_{s\to r_1}[q_{11}^{-1}(s)(s+r_1)] \\ R_2 &= \lim_{s\to r_2}[q_{11}^{-1}(s)(s+r_2)] \end{aligned} \right\} \tag{5.13}$$

or in the time domain

$$P_1(t) = R_1 e^{-r_1 t} + R_2 e^{-r_2 t}. \tag{5.14}$$

The other state probabilities can be derived from Q_s^{-1} in a similar manner. Table 5.1 shows some important Laplace transforms often encountered in reliability analysis.

5.3.2 Evaluation of mean and steady-state characteristics

When both transient characteristics and non-transient characteristics have to be found it is best to solve the differential equations and to use the resulting time-dependent state probabilities to calculate the appropriate mean or steady-state values. However, when no time-dependent values have to be evaluated the use of differential equations for state-space model description can be avoided. This section will show the various ways to solve the time-independent quantities.

Table 5.1 Laplace transforms

Time domain	*Laplace domain*
e^{-at}	$\frac{1}{s+a}$
$\frac{t^n}{n!}e^{-at}$	$\frac{1}{(s+a)^{n+1}}$
$\int f(t)\mathrm{d}t$	$\frac{F(s)}{s}$
$\frac{\mathrm{d}f(t)}{\mathrm{d}t}$	$sF(s) - f(0)$
$\lim_{t\to 0} f(t)$	$\lim_{t\to\infty} sF(s)$
$\lim_{t\to\infty} f(t)$	$\lim_{s\to 0} sF(s)$
C	$\frac{C}{s} \quad (s > 0)$

(a) *Steady-state characteristics*

When the time-dependent state probabilities are available, the steady-state characteristics A_∞ (steady-state availability), MTBF (mean up-time between failures), MTTR (mean time to repair) and τ_c (mean cycle time) can be calculated from equations (5.7) and (5.8) as

$$\left.\begin{aligned} A_\infty &= \lim_{t\to\infty} A(t) \\ MTBF &= \frac{1}{\lambda_s(\infty)} = \lim_{t\to\infty}\frac{1}{\lambda_s(t)} \\ \mathrm{MTTR} &= \frac{1}{\mu_s(\infty)} = \lim_{t\to\infty}\frac{1}{\mu_s(t)} \\ \tau_c = \mathrm{MTTR} + \mathrm{MTBF} &= \frac{\mathrm{MTBF}}{A_\infty} = \frac{\mathrm{MTTR}}{\bar{A}_\infty}. \end{aligned}\right\} \tag{5.15}$$

However, for repairable systems with a healthy repair strategy the transient in the $A(t)$ function will last only a short time, say three times the MTBF, i.e. the $A(t)$ function will quickly settle at its steady-state value. Therefore, in practice, one is usually interested in the steady-state characteristics only. In the steady-state situation the derivatives of the state probabilities will be zero and consequently the differential equations can be rewritten as linear equations like

$$\acute{P}_i(\infty) = 0 = \sum_{\substack{k=1\\k\neq i}}^{n} P_k(\infty)z_{ki} - \sum_{\substack{k=1\\k\neq i}}^{n} P_i(\infty)z_{ik} \quad (i = 1, 2, \ldots, n) \tag{5.16}$$

or in matrix notation

$$\left.\begin{aligned}[\acute{P}_i(t)] = [0] = [P_i(t)]Q_t \\ \text{or} \qquad PQ_t = [0]\end{aligned}\right\} \tag{5.17}$$

However, in zeroing the state probabilities' derivatives some information is lost which effectively reduces the number of independent equations to $n - 1$. Since there are n variables to be solved we have to replace one of the equations by the extra equation

$$\sum_{i=1}^{n} P_i = 1 \tag{5.18}$$

to obtain a solvable set of equations again. A very simple way to do so is to replace one column of Q_t in equation (5.17) with a column of 1s (yielding $Q_{t,1}$) and simultaneously to replace the corresponding element of the $[\acute{P}]$ vector with a 1 as well. With these adaptions to equation (5.17) the vector of state probabilities can be found as the corresponding row of the inverse matrix $(Q_{t,1})^{-1}$. After solving the state probabilities the steady-state characteristics can be evaluated using formulas (5.7), (5.8) and (5.15).

In system design one often has the choice between a single unit and one or more redundant configurations of identical items. Then for ease of comparison it is convenient when the availability of these redundant systems can be expressed in terms of the availability of the single unit. This can be done simply by rewriting the transition matrix Q_t with a for μ and $(1 - a)$ for λ where a is the unit availability $\mu/(\mu + \lambda)$. For example for a one-out-of-two standby system with two repair teams this yields the matrix

$$\begin{aligned} Q_m = \frac{1}{\lambda + \mu} Q_t &= \frac{1}{\lambda + \mu}\begin{bmatrix} -\lambda & \lambda & 0 \\ \mu & -\mu-\lambda & \lambda \\ 0 & 2\mu & -2\mu \end{bmatrix} \\ &= \begin{bmatrix} -(1-a) & 1-a & 0 \\ a & -1 & 1-a \\ 0 & 2a & -2a \end{bmatrix} \end{aligned} \tag{5.19}$$

When we replace the second column of Q_m by 1s the state probabilities can be found as the elements on the second row of the inverse matrix $(Q_{m,1})^{-1}$, i.e.

$$\begin{aligned} Q_{m,1}^{-1} &= \begin{bmatrix} -(1-a) & 1 & 0 \\ a & 1 & 1-a \\ 0 & 1 & -2a \end{bmatrix}^{-1} \\ &= \frac{1}{(1+a)^2}\begin{bmatrix} \cdots & \cdots & \cdots \\ 2a^2 & 2a(1-a) & (1-a)^2 \\ \cdots & \cdots & \cdots \end{bmatrix}. \end{aligned} \tag{5.20}$$

(b) *Mean time to first failure (MTTFF)*

When the reliability has to be found as well, the MTTFF can be derived in two manners: as the time integral of the $R(t)$ function over the interval $[0,\infty]$:

$$\text{MTTFF} = \lim_{T\to\infty}\int_0^T R(t)\mathrm{d}t = \sum_{i\in U}\left(\lim_{T\to\infty}\int_0^T P_i(t)\mathrm{d}t\right) \tag{5.21}$$

or as the zero value of the reliability function in the Laplace domain

$$\text{MTTFF} = \sum_{i\in U}\left(\lim_{s\downarrow 0} P_i(s)\right) \tag{5.22}$$

which results from equation 5.21 through application of the initial-final value theorem. This latter manner avoids the inverse Laplace transform.

However, the Laplace transformation can be avoided completely when we integrate both the left-hand side and the right-hand side of the original differential equation

$$\int_0^\infty \frac{\mathrm{d}P_i(t)}{\mathrm{d}t}\mathrm{d}t = P_i(\infty) - P_i(0) = \sum_{\substack{j=1\\ j\neq i}}^{n}\Theta_k z_{ki} - \sum_{\substack{k=1\\ k\neq i}}^{n}\Theta_i z_{ik}$$

$$(i = 1, 2, \ldots, n)$$

with

$$\Theta_i = \int_0^\infty P_i(t)\mathrm{d}t. \tag{5.23}$$

The only non-zero initial condition is $P_1(0) = 1$ for the starting state. Since the absorbing down state does not affect the MTTFF, the set of equations can be reduced by the restriction $i, k \in U$. By solving the remaining set of linear equations the MTTFF can be calculated easily as

$$\text{MTTFF} = \sum_{i\in U}\Theta_i. \tag{5.24}$$

EXAMPLE 5.3

For the redundant hospital power-supply system with identical failure and repair rates for both the public power supply and the emergency power supply as discussed in Example 5.2 the set of equations for direct MTTFF evaluation is

$$[\Theta_1\ \Theta_2]\begin{bmatrix} -2\lambda & 2\lambda \\ \mu & -(\lambda+\mu)\end{bmatrix} = [-1\ 0]. \tag{5.25}$$

Then, the mean time to failure for this system is calculated as

$$\text{MTTFF} = \Theta_1 + \Theta_2 = \frac{\lambda+\mu}{2\lambda^2} + \frac{1}{\lambda} = \frac{3\lambda+\mu}{2\lambda^2}. \tag{5.26}$$

5.4 APPROXIMATED SOLUTION METHODS USING FINITE TIME STEPS

The Laplace solution method introduced in the previous paragraphs leads to exact analytical solutions of homogeneous Markov problems, however the larger the Markov problem (diagram) the more difficult analytic solutions become. Therefore, for larger problems one usually adopts numerical methods. In these methods the differential equations (5.1) are approximated by difference relations where transitions are governed by conditional transfer probabilities in finite time intervals instead of transfer rates

$$\left[\frac{P_i(t + \Delta t) - P_i(t)}{\Delta t}\right] = [P_i(t)]Q_t. \tag{5.27}$$

This can be rewritten as

$$P(t + \Delta t) = P(t)(I + Q_t\Delta_t) \approx P(t)Q_p. \tag{5.28}$$

Here, Q_p is the matrix of conditional probabilities $q_{p,ij}$ that describe the probability the system will be in state S_j at $t + \Delta t$ given the condition that the system was in state S_i at time t. Then the system state probabilities at each consecutive interval Δt can be evaluated as:

$$P_n = P(n\Delta t) = P_0(Q_p)^n. \tag{5.29}$$

The matrix $(I + Q_t\Delta t)$ is a first-order approximation of Q_p and is legal only when the time steps Δt are considerably smaller than all mean transition times ($1/\lambda$s and $1/\mu$s) in the system. A more accurate approximation is achieved when the failures and repairs of items are approximated as $f = 1 - e^{-\lambda\Delta t}$ and $r = 1 - e^{-\mu\Delta t}$, respectively. Then the elements $q_{p,ii}$ and $q_{p,ij}$ of the probability matrix can be found as the probability that no transition from state i occurs in Δt ($q_{p,ii}$) and the probability that a transition from state i to (another) state j occurs in Δt given the condition that no other transitions from this state have occurred ($q_{p,ij}$), respectively. In formula this can be written as

$$\begin{aligned} q_{p_{ii}} &= \prod_k e^{-\lambda_{ik}\Delta t} = \prod_k (1 - f_{ik}) \\ q_{p_{ij}} &= (1 - e^{-\lambda_{ij}\Delta t})\prod_{k\neq j} e^{-\lambda_{ik}\Delta t} = f_{ij}\prod_{k\neq j}(1 - f_{ik}). \end{aligned} \tag{5.30}$$

Although the conditional probabilities evaluated in this manner are more accurate than those according equation (5.28) these are still approximations and do not take account of the possibility of multiple events (transitions) in a single interval Δt. Therefore, the time interval Δt should be chosen small enough with respect to the shortest mean transition times $1/\lambda_{ij}$, which usually are the mean repair times of components.

5.5 CONCLUSIONS

The state-space method is readily applicable to systems having a finite number of states and allows multi-state modelling for both the system and its components. Usually, analytical solutions only exist when all transition rates are constant. Even then manually solution becomes very complex when more than three states are involved. Therefore, calculations are usually performed with help of computer programs that use the state-space diagram and transition rates as inputs.

For systems with time-dependent transition rates both analytical methods and numerical methods are generally insufficient and statistical failure data can only be determined with help of Monte-Carlo simulation methods. However, in some cases the lifetime of an item can be described as the sum of several negative exponentially distributed lifetimes. These special distributions can be incorporated into a Markov model with merely constant transition rates by introduction of so-called dummy states. One example of such a distribution is the n-Erlang distribution which can be described by a sequence of n transitions with identical failure rates.

5.6 WORKED PROBLEMS

PROBLEM 5.1

A two-out-of-three standby redundant system consists of three identical units of which only two are required to run. The other one is either held in reserve (standby) or under repair following failure. When only one unit is left operational it is useless on its own and, therefore, it is stopped to prevent further failures. Only one repair man is present who starts as soon as the first unit fails and continues until all units are available again. The system is restarted after failure as soon as two units are ready for use. The failure and repair rates of the units are λ and μ, respectively.

1. Construct the Markov diagram for this system.
2. Calculate the MTTF.

ANSWER

1.

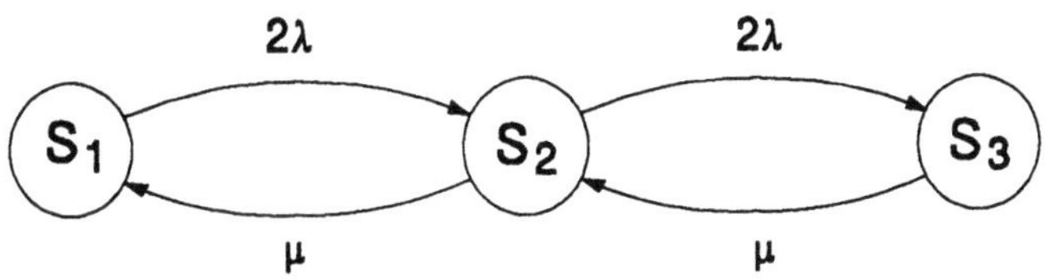

Fig. 5.6 Markov diagram for Problem 5.1.

2. The mean times θ_i spent by the system in each up-state before failure can be found as the elements of the first row of the inverse transition matrix

$$\theta = [-1\ \ 0]\begin{bmatrix} -2\lambda & 2\lambda \\ \mu & -(2\lambda+\mu) \end{bmatrix}^{-1}$$

$$= [-1\ \ 0]\begin{bmatrix} \dfrac{-(2\lambda+\mu)}{4\lambda^2} & -\dfrac{1}{2\lambda} \\ \dfrac{\mu}{4\lambda^2} & \dfrac{-1}{2\lambda} \end{bmatrix} = \begin{bmatrix} \dfrac{2\lambda+\mu}{4\lambda^2} & \dfrac{1}{2\lambda} \end{bmatrix}. \qquad (5.31)$$

Then the mean time to failure can be found simply as the sum of all θ_is as

$$\text{MTTF} = \theta = \frac{4\lambda+\mu}{4\lambda^2}. \qquad (5.32)$$

PROBLEM 5.2

A polar expedition has two identical motor sleighs at its disposal for transportation purposes. Each sleigh is built up of six different replaceable components all having a failure rate of 0.05 a month under normal use. When only one sleigh is left available this will be used more frequently and the effective component failure rates become 0.1 a month. Broken components cannot be repaired and, therefore, for each type of component one spare part is taken with. Considering that the occurrence of all component failures is statistically independent and that spare parts in storage cannot fail, calculate the probability that after a six-month stay at least one sleigh is still available.

ANSWER

For each type of component the following Markov diagram can be drawn.

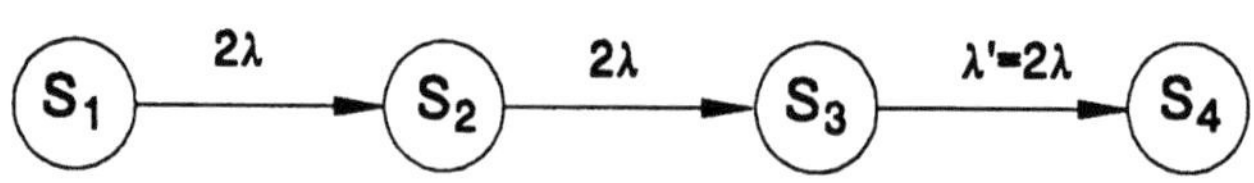

Fig. 5.7 Markov diagram for Problem 5.2.

The state probabilities (in Laplace notation) can be found as the elements on the first row of the inverse matrix Q_s^{-1}

$$[P_i(s)] = [1\ \ 0\ \ 0]\begin{bmatrix} s+2\lambda & -2\lambda & 0 & 0 \\ 0 & s+2\lambda & -2\lambda & 0 \\ 0 & 0 & s+2\lambda & -2\lambda \\ 0 & 0 & 0 & s \end{bmatrix}^{-1}$$

$$= \left[\frac{1}{s+2\lambda}\ \frac{2\lambda}{(s+2\lambda)^2}\ \frac{(2\lambda)^2}{(s+2\lambda)^3}\ \frac{(2\lambda)^3}{(s+2\lambda)^3 s}\right]. \tag{5.33}$$

Then $R(s)$ can be written in sepearate terms as

$$R(s) = \sum_{i=1}^{3} P_i(s) = \frac{1}{s+2\lambda} + \frac{2\lambda}{(s+2\lambda)^2} = \frac{(2\lambda)^2}{(s+2\lambda)^3}, \tag{5.34}$$

which translates back to the time domain as

$$R(t) = \mathrm{e}^{-2\lambda t}\left[1 + 2\lambda t + \frac{(2\lambda t)^2}{2!}\right]. \tag{5.35}$$

The probability that at least one item of each type is available after six months is then

$$R_s(t=6) = R(t=6)^6 \approx 0.869. \tag{5.36}$$

PROBLEM 5.3

The machinery of a workshop consists of a lathe and a milling machine for which only one repair man is available. Repair starts as soon as the first failure is reported and continues until both machines are available again. The repair strategy is such that the machines are repaired in order of failure. The machines fail statistically independent with failure and repair rates

$$\lambda_l = 1/\text{year},\ \mu_l = 100/\text{year for the lathe.}$$

and

$$\lambda_m = 2/\text{year},\ \mu_m = 100/\text{year for the milling machine.}$$

1. Construct the Markov diagram for this machinery and repair strategy.
2. Calculate the repair man's degree of occupation, i.e. the probability that the repair man is busy.
3. Calculate the steady-state availability and the mean down time (MTTR) of the lathe.

ANSWER

1. The Markov diagram for this problem is shown in Fig. 5.8.
2. The transition matrix Q_t can be found from the Markov diagram as

$$Q_t = \begin{bmatrix} -(\lambda_l+\lambda_m) & \lambda_l & \lambda_m & 0 & 0 \\ \mu_l & -(\lambda_m+\mu_l) & 0 & \lambda_m & 0 \\ \mu_m & 0 & -(\lambda_l+\mu_m) & 0 & +\lambda_l \\ 0 & 0 & \mu_l & -\mu_l & 0 \\ 0 & \mu_m & 0 & 0 & -\mu_m \end{bmatrix} \tag{5.37}$$

The steady-state probabilities can be found by exchanging one of the columns of Q_t (e.g. column one) with a column of 1s thus creating

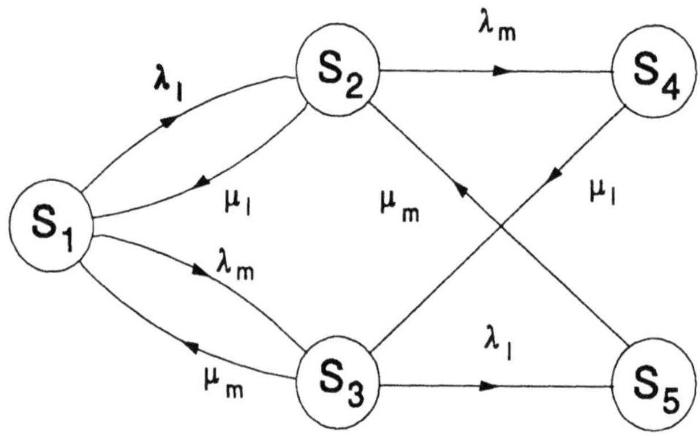

Fig. 5.8 Markov diagram for Problem 5.3.

a new matrix $Q_{t,1}$. Then the state probabilities are found as the corresponding row (in this case row one) of the inverse of this matrix

$$[P_1\ P_2\ P_3\ P_4\ P_5] = \begin{bmatrix} \frac{625}{644} & \frac{25}{2576} & \frac{25}{1288} & \frac{1}{5152} & \frac{1}{5152} \end{bmatrix}. \tag{5.38}$$

The degree of occupation is then: $1 - P_1 = 19/644 = 0.0295$.

3. The availability of the lathe is $A_l = P_1 + P_2 = 1275/1288 = 0.9899$, while the effective repair time of the lathe can be found as

$$\phi = \frac{1}{\lambda_l}\frac{1 - A_l}{A_l}. \tag{5.39}$$

PROBLEM 5.4

A patrol vessel of the coast guard has two identical engines each of which can supply enough power to allow sailing at cruise speed. In normal use both engines are active and have a failure rate $\lambda_e = 10^{-3}$/h. At failure of the first engine the other one has to supply the full load which increases the failure rate to $\lambda_e' = 3 \times 10^{-3}$/h. The propulsion system fails when both engines fail.

1. Evaluate the MTTF (θ) when no corrective maintenance is applied to a single failure.
2. What does the MTTF become when an (on-board) repair with repair rate $\mu_e = 10^{-1}$ is started as soon as an engine fails?
3. Evaluate the frequency of repair calls for case 2.
4. Repair of an engine costs £500 on average. When, however, the complete propulsion system fails the costs will be raised by £4500 due to loss of mission. Decide which of the repair strategies is favourable and justify your answer.

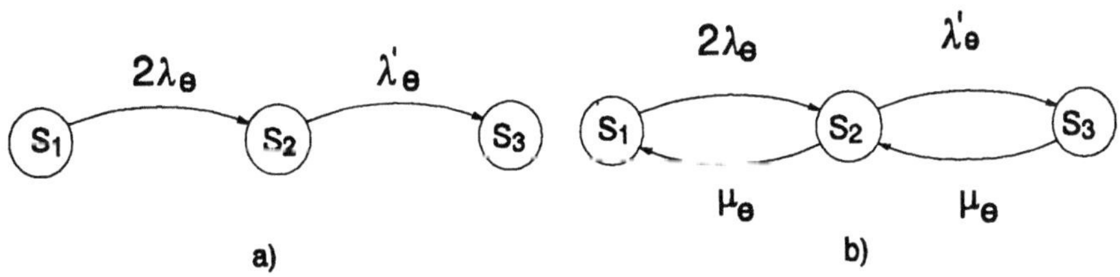

Fig. 5.9 Markov diagrams for Problem 5.4.

ANSWER

1. From Fig. 5.9(a) the MTTF without repair can be evaluated as

$$\mathrm{MTTF_a} = \frac{1}{2\lambda_e} + \frac{1}{\lambda'_e} = 833\tfrac{1}{3}\,\mathrm{h}. \tag{5.40}$$

2. With maintenance the Markov diagram of Fig. 5.9(b) applies. The mean times θ_i spent by the system in each up-state before failure can be found as the elements of the last row of the inverse transition matrix minus the elements on the first row of this matrix.

   Then the mean time to failure can be found simply as the sum of all θ_is as

$$\theta = [-1\ \ 0]\begin{bmatrix} -2\lambda_e & 2\lambda_e \\ \mu_e & -(\lambda'_e + \mu_e) \end{bmatrix}^{-1}$$

$$= [-1\ \ 0]\begin{bmatrix} \dfrac{-(\lambda'_e + \mu_e)}{2\lambda_e\lambda'_e} & -\dfrac{1}{\lambda'_e} \\ \dfrac{-\mu_e}{2\lambda_e\lambda'_e} & -\dfrac{1}{\lambda'_e} \end{bmatrix} = \begin{bmatrix} \dfrac{\lambda'_e + \mu_e}{2\lambda_e\lambda'_e} & \dfrac{1}{\lambda'_e} \end{bmatrix}. \tag{5.41}$$

$$\mathrm{MTTF_b} = \theta = \frac{2\lambda_e + \lambda'_e + \mu_e}{2\lambda_e\lambda'_e} = 17\,500\,\mathrm{h}. \tag{5.42}$$

3. Under the condition that $\lambda_e, \lambda'_e \ll \mu_e$ the frequency of repair calls is dominated by the cycle time of the first loop in the diagram and, therefore, can be approximated as

$$f = \frac{a}{\dfrac{1}{2\lambda_e} + \dfrac{1}{\mu_e}} = 1.96 \times 10^{-3}\,\mathrm{h}^{-1}. \tag{5.43}$$

4. The average repair costs per unit of time are 5500/$\mathrm{MTTF_a}$ = 6.60 £/h when repair starts after system failure and ($f \times 500 + 5000$)/$\mathrm{MTTF_b}$ = 0.28 £/h when repair starts after unit failure. It will be obvious that repair strategy 2 is the best choice.

REFERENCES

Dhillon, B.S. and Singh, C. (1981) *Engineering Reliability – New Techniques and Applications*, Wiley, New York.

Kapur, K.C., *et al.* (1977) *Reliability in Engineering Design*, Wiley, New York.

Klaassen, K.B. and van Peppen, J.C.L. (1988) *System Reliability, Concepts and Applications*, Edward Arnold, London.

Pages, A. and Gondran, M. (1986) *System Reliability, Evolution and Prediction in Engineering*, North Oxford Academic, London.

Smith, D.J. (1972) *Reliability Engineering*, Pitman, London.

CHAPTER 6

Productiveness of manufacturing systems

'Productiveness' is a word we have invented for the concept of expected output rate of a fallible, repairable, productive system. As used in this book, productiveness is defined as the actual average production rate of a system when operated under prescribed physical and managerial conditions expressed as a fraction of its maximum output rate when all items in the system are operating or capable of operating correctly. Productiveness is thus the appropriate measure of comparative system effectiveness for productive systems, that is, it can be used to compare systems and to assess improvements to existing systems. In systems in which the quality of the product varies with the state of the system it may be necessary to qualify this definition. We have deliberately avoided use of the word 'productivity' because it means different things to different people.

The effectiveness of other types of system has traditionally been measured in ways which differ somewhat from the analysis given in this chapter. In weapon systems, for example, effectiveness is a combination of the operational availability of the delivery system, the reliability of the weapon itself and the less-than-perfect accuracy on target inherent in a properly functioning weapon given a properly functioning guidance system handled by an average operator. Such problems amount to no more than the correct combination of probabilities and do not involve new concepts. The productive system, however, does require some new ideas. The first of these is acceptance that neither availability nor reliability is a satisfactory comparator for a production manager or plant engineer seeking to assess competing productive systems.

We do not consider, in this analysis, the logistics of spare parts supply. Productiveness and costs are affected by the policies of both the supplying and the using company, because long delays may occur if spares are not to hand when needed and holding spares costs money. Such matters are beyond the scope of this book, however.

We consider a generalized system of N stages in series in which each stage consists of n_i identical items each with infinitely variable output up to a maximum value. Analytical approximations are derived for the case where there is no interstage storage (i.e. minimum work in progress).

This case has recently become more realistic with the introduction of just-in-time (JIT) manufacturing and is approached quite closely in factories using the more sophisticated manufacturing resource planning (MRPII) computer systems in conjunction with computer integrated manufacturing (CIM).

6.1 EFFECTS OF JIT/KANBAN

Under the JIT manufacturing system, nothing is made and nothing ordered until there is only just time to make it or obtain it. When a downstream activity requires more of a certain part it sends a KANBAN (i.e. a reusable order token) to the section which makes the component for a limited number of parts to be made, ideally just one, but usually up to about 100 and typically 40. Production is therefore 'pulled' through the system by the customers rather than 'pushed' from the supply end. Effectively, there are no parts in interstage storage on a JIT production line. There is no provision for either defective parts or machinery breakdown in such a system. We regard this as somewhat unrealistic. Interstage storage is really an economic alternative to redundancy or more reliable plant which needs to be calculated rather than shunned on 'ideological' grounds. Some processes are inherently marginal for quality control and are bound to produce some defects, e.g. the manufacture of silicon chips.

Advocates of JIT claim that it leads to more efficient production with much lower scrap and rework rates and that it instils a sense of responsibility in all workers. In practice, it is coupled with total productive maintenance (TPM), another Japanese method which involves the production workers in keeping the machinery in good condition. This is necessary because JIT does not allow PM to be scheduled other than in the dead time arising at weekends, etc. Rather, the system gives rise to ripples of unscheduled free time which travel both up and down the production line from the site of a plant breakdown or other delay. If these opportunities for PM are not used, system condition deteriorates and all the gains and more from using JIT are lost through unreliability.

It seems to us better and more scientific to try to achieve an optimization of plant provision, manning and maintenance through a modern MRPII system. Such systems use a computer to preplan production and can be made to allow for plant maintenance as another activity involving the machines in question. MRPII can show the need for redundancy and calculate the economic batch quantities.

Referring back to the discussion of the bathtub curve in Chapter 1, production planners should understand that the ROCOFs and repair rates of complex systems depend upon the conscientious and regular completion of preventive maintenance (PM). The optimization of schedules is alluded to in Chapter 11 but is really a subject for another book. Suffice it to say here that PM should not be regarded as optional because, if it is properly applied, it minimizes the cost of downtime from PM and failures combined. If it is omitted, the ROCOF will rise. It is relatively easy to add on the effect of scheduled PM within the MRPII system although there is some doubt about how to make such a system prioritise PM correctly. If PM is regarded simply as an expense, it will get deferred when the facility is busy as an activity producing no or negative profit. A way needs to be found to condition the ROCOF and repair costs upon the completion of the PM.

In this chapter we derive methods, which can be used in an MRPII system, for predicting the mean rate of production from a given system and the effects of various changes to its configuration. Analytical results are feasible only when there is no interstage storage, which is the aim in both JIT and MRPII systems. Simulation features in the latest MPRII systems are able to simulate the effects of interstage storage.

6.2 PRODUCTIVENESS OF SYSTEMS WITH NO INTERSTAGE STORAGE

6.2.1 Review of basics from a productiveness point of view

Recall from Chapter 2 that the steady state availability of a series system is given by

$$A_s = 1\bigg/\left[1 + \sum_{i=1}^{N}((1 - A_i)/A_i)\right] \tag{6.1}$$

for a system of N stages in series having availability A_i, $i = 1, 2, \ldots, N$. It was asserted in Chapter 2 that this result is non-parametric. It follows that if, say, the ith stage consists of a number of items, with, possibly, some redundancy, the stage result can still be used in equation (6.1) without error. It also follows that if we change the definition of system available from 'capable of producing at $100\,V_1\%$ of maximum output' to 'capable of producing at $100\,V_2\%$ of maximum output' then equation (6.1) is again still valid but with changed values for the A_i.

For $N = 2$ equation (6.1) becomes

$$A_s = A_1A_2/(A_1 + A_2 - A_1A_2). \tag{6.2}$$

It follows that if an additional stage numbered $N + 1$ is added to an existing series system, equation (6.1) becomes

$$A_s^{(N+1)} = A_s^{(N)}A_{N+1}/(A_s^{(N)} + A_{N+1} - A_s^{(N)}A_{N+1}). \tag{6.3}$$

6.2.2 Stage operating policies

Redundant stages can be dichotomously classified in four independent ways. In productive systems, the sum of these classifications is often determined by the operating policy as well as the physical arrangement of components.

1. *Active parallel or standby.* Active redundant elements are at risk before they are needed; standbys are not. It is usually assumed, as it is here, that reduction of stress by active parallel operation does not affect ROCOF.

2. *Maintained or unmaintained.* In a maintained stage, repairmen work on failed items while other items sustain stage function. In unmaintained stages, no repairs (can) start until stage failure, which implies system failure. All items are then repaired. (Any other policy would amount to permanent reduction of redundancy, except for opportunity repairs during system failures due to other stages.)

3. *Limited and unlimited repairs.* In general, $r \leqslant n - m + 1$ simultaneous repairs are possible. In the limited case, $r < n - m + 1$. In the case of stages which continue on system failure (see below), $r < n$ or $r = n$, respectively.

4. *Stop or continue on stage failure.* Just as redundancy itself is often a matter of operating policy rather than material arrangement, so stage failure is subject to definition. A stage failed by definition may have some operable items which can do useful work.

6.2.3 Redundancy in productive systems

In productive systems, stages may have several productive states distinguished by differences in output rate and/or (less often) quality. In calculating the productiveness and expected profit or benefit from a system, all the possible combinations of productive stage states and the proportions of total time over which they are expected to exist must be known. The availability of such systems is usually defined, as it is here, as the probability that the normal production rate is possible at any instant. This normal production rate may be lower than the absolute maximum instantaneous rate, if some redundancy is created by operating policy, as opposed to occurring consequential to system configuration.

6.2.4 Active parallel availability

Maintained stages in active redundancy with ulimited repairs and which *continue on stage failure* follow the same binomial algebra as the reliability case, i.e. rewriting equation (3.8)

$$\hat{A}_i = \sum_{j=0}^{n-m}\left[\binom{n}{j} a_i^{n-j}(1 - a_i)^j\right], \tag{6.4}$$

where $\hat{A}_i$ is the stage availability of an m-out-of-n parallel stage with identical items of availability a_i.

If survivors are stopped during stage failure and $m > 1$, then the same problem arises as in the series case ($m = n$ *is* the series case!). The probability of states below the just-failed state is zero, in fact, but greater than zero by implication in equation (6.4). This can be adjusted by reducing the total probability-space by p_o, where

$$p_o = \sum_{k=0}^{m-2}\left[\binom{n}{k} a_i^k(1 - a_i)^{n-k}\right]. \tag{6.5}$$

Then, as we saw in equation (3.9),

$$A_i = \hat{A}_i/(1 - p_o). \tag{6.6}$$

6.2.5 Standby availability

For the convenience of readers, some of the more common results for standby stages, calculated by the matrix method of Chapter 5, are given below.

Maintained standby, $n = 2$, $m = 1$.

$$A = 2a/(1 + a^2), \quad r = 2; \tag{6.7}$$

$$A = a/(1 - a + a^2), \quad r = 1; \tag{6.8}$$

$$A = 1 - \frac{\lambda_A\lambda_B(\lambda_A + \mu_B)}{\mu_B(\mu_A + \lambda_b)(\lambda_A + \mu_B) + \lambda_A\lambda_B(\mu_A + \mu_B + \lambda_A)}, \tag{6.9}$$

$r = 1$, item A preferred;

$$A = 2aS/(1 - a + aS + a^2S), \quad r = 2. \tag{6.10}$$

In (6.10) switching is successful with probability S, with a common repair rate for item and switching failures. Alternatively $(1 - S)$ could be the probability of an unrevealed fault in the standby which becomes apparent when it is switched on.

Maintained standby, $n = 3$, $m = 2$, stop on stage failure:

$$A = (2a - a^2)/(2 - 2a + a^2), \quad r = 2; \tag{6.11}$$

$$A = (2a - a^2)/(4 - 6a + 3a^2), \quad r = 1. \tag{6.12}$$

Maintained standby, $n = 3$, $m = 2$, continue on stage failure:

$$A = 3a^2(2 - a)/(a^3 + 2), \quad r = 3; \tag{6.13}$$

$$A = (2a^2 - a^3)/(1 - a + a^2), \quad r = 2; \tag{6.14}$$

$$A = (2a^2 - a^3)/(4 - 8a + 6a^2 - a^3), \quad r = 1. \tag{6.15}$$

Maintained standby, $n = 3$, $m = 1$:

$$A = 3a(1 + a^2)/(2a^3 + 3a + 1), \quad r = 3; \tag{6.16}$$

$$A = 2a(1 + a^2)/(a^3 + 3a^2 - a + 1), \quad r = 2; \tag{6.17}$$

$$A = a(a^2 - a + 1)/(2a^2 - 2a + 1), \quad r = 1. \tag{6.18}$$

6.2.6 Unmaintained standby stages

For constant μ and λ, with perfect switching, the quickest calculation of availability is through stage MTBF and MTTR. The matrix method is valid, but in constructing Q, allowance must be made in the downstates for the completion of each successive repair. For example, the matrix for a two-out-of-four unmaintained standby system with two repairmen, would be

$$Q = \begin{bmatrix} -2\lambda & 2\lambda & 0 & 0 & 0 & 0 \\ 0 & -2\lambda & 2\lambda & 0 & 0 & 0 \\ 0 & 0 & -2\lambda & 0 & 0 & 2\lambda \\ \mu & 0 & 0 & -\mu & 0 & 0 \\ 0 & 0 & 0 & 2\mu & -2\mu & 0 \\ 0 & 0 & 0 & 0 & 2\mu & -2\mu \end{bmatrix}. \tag{6.19}$$

Even without the matrix it is obvious that in this example the MTBF is

$$\theta = 3/2\lambda \tag{6.20}$$

and in general, for unmaintained stages,

$$\theta = (n - m + 1)/m\lambda. \tag{6.21}$$

The example MTTR is clearly

$$\phi = 2 \times 1/2\mu + 1/\mu = 2/\mu \tag{6.22}$$

and, in general,

$$\phi = r(n - m + 2 - r)/\mu + \sum_{i=1}^{r-1} i/u. \tag{6.23}$$

Then, as usual,

$$A = \theta/(\theta + \phi). \tag{6.24}$$

θ is a multiple of $1/\lambda$ and ϕ is a multiple of $1/\mu$, and so the availability can always be expressed in terms of $\rho = \mu/\lambda$ or $a = \mu/(\mu + \lambda)$. In the two-out-of-four example $r = 2$

$$\begin{aligned} A &= 1/(1 + 4/3\rho) \\ &= 3a/(4 - a). \end{aligned} \tag{6.25}$$

6.2.7 Generality of matrix approach

The matrix approach through simultaneous equations succeeds for any stage or system whose behaviour can be expressed by means of a state transition rate matrix with constant coefficients. If, additionally, there are only two different constant rates in the matrix, then all the matrix equations can be expressed in terms of a single variable, the ratio of the original pair. In particular, if the items represented by the matrix are identical then the two constants μ and λ can be reduced to one, ρ, in all the matrix equations. Therefore A can be found in terms of a for all stages of identical items with constant μ and λ, provided that either ρ or a is known; it is not necessary to know μ and λ separately.

6.2.8 Productiveness of a single stage

The productiveness of a single stage is the probability-sum of the output rates for every state of the stage, relative to stage-rated output

$$E_i \equiv \sum_{j=1}^{n} (p_j v_j). \tag{6.26}$$

For a stage where v_j is the proportion of rated output achieved in state j and p_j is the probability of state j with constant μ and λ and identical items, equation (6.26) can be written as a vector product

$$E_i \equiv \mathbf{P}.\mathbf{V}. \tag{6.27}$$

If $v_{i,j}$ is the relative output of stage i in state j, then $v_{i,1} \equiv 1$ for all i, that is, no stage is ever operated above the system-rated maximum output.

6.2.9 System productiveness

The simple product of the stage productiveness underestimates the system productiveness. A better approximation, developed below, takes account of the probability of running with more than one stage at reduced output. It does *not* account for deliberate restriction of risk of failure by shutting down unneeded items when output is reduced by failures in other stages. The calculation is therefore conservative, except for the case of an operating policy in which items are slowed, rather than selectively stopped, to rebalance the line following failures.

Define a vector $\boldsymbol{W}$ with elements $w_{k,h}$ which are the sums of the probabilities of the h greatest production rates of stage k, i.e.

$$w_{k,h} \equiv \sum_{j=1}^{h} p_{k,j}. \tag{6.28}$$

Then the probability that the system output is $v_{i,j}$ (i.e. the jth greatest output rate of stage i) is approximated by

$$Pr(v_{i,j}) \simeq p_{i,j} \prod_{k \neq i} w_{k,h}. \tag{6.29}$$

By applying the logic of the series availability calculation (6.1), a more precise evaluation is

$$Pr(v_{i,j}) = p_{i,j} \Big/ \left[1 + \sum_{k \neq i} ((1 - w_{k,h})/w_{k,h}) \right]. \tag{6.30}$$

Both equations (6.29) and (6.30) are subject to

$$v_{k,h} \geqslant v_{i,j} > v_{k,h+1}, \quad \text{for } i < k \tag{6.31}$$

and

$$v_{k,h} > v_{i,j} > v_{k,h+1}, \quad \text{for } i > k. \tag{6.32}$$

Note that condition (6.32) still does not account for the reduction of risk produced by shutting down unneeded machines during restricted production due to failures.

The system productiveness is then

$$E = \sum_{i} \sum_{j} (v_{i,j} Pr(v_{i,j})). \tag{6.33}$$

In evaluating equations (6.30) and (6.33), it is necessary to avoid double counting, where $v_{k,h} = v_{i,j}$. In coding for computers, it is quite simple to arrange that such equality never occurs by making equal values differ on input by negligible amounts, which are nevertheless discernible to the processor. It is more difficult, but not impossible, to code the actual conditions given, i.e. to allow the equality for $i < k$ but to disallow it for $i > k$.

6.3 EFFECT OF INTERSTAGE STORAGE – SIMULATION

The case of no interstage storage which we examined above represents one extreme position. At the other extreme is the case where there is a store of infinite capacity between each pair of adjacent stages. In this case it is immediately obvious that the productiveness of the system is equal to that of the least productive stage. The system dynamics in the general case of finite stores are impossible to model analytically without making assumptions, such as perfect 'balance' between stages, i.e. all v_i equal, which are not universally applicable. Luckily, computer programs for simulation, e.g. MIRIAM from Scicon, are available which can solve such systems for mean and variance of output rate and many other factors. MIRIAM is based upon 'next event' simulation; that is all the distributions of failure and repair times are known and random numbers are generated to decide when each will fail next. The 'next event' is the earliest of these simulated failures. This is preferable to simulation of the 'ticking clock' type in which simulated time is advanced by one tick, e.g. an hour, at a time and the state of the system for that tick determined from present probabilities.

Ticking clock simulations take more computer time which is no longer very important, but the software programs through which they operate generate statistical outputs based upon operating time rather than number of events. The accuracy of the simulation depends on the number of events rather than the time simulated. As a bare minimum, every item should fail at least once in a simulation run.

6.4 WORKED PROBLEMS

PROBLEM 6.1

Use the methods of Chapter 5 to prove some of the equations for the availability of maintained standby systems.

ANSWER

The methodology is the same for all so two examples will suffice

1. $$n = 2, m = 1, r = 1.$$

First define the states of the system:

S_1 both available one running;
S_2 one running one under repair;
S_3 both failed one under repair.

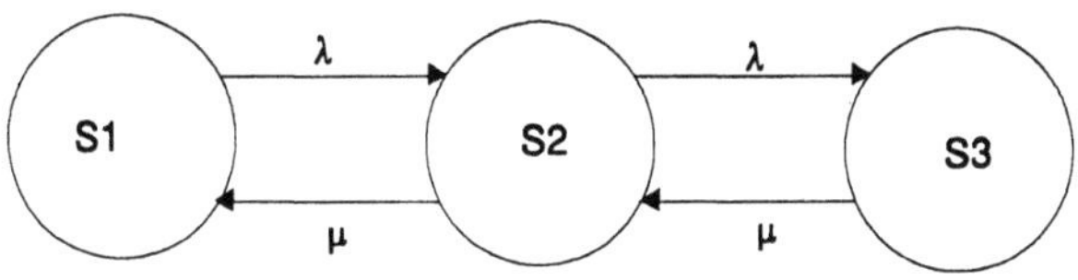

Fig. 6.1 State-space diagram for Problem 6.1.

Then draw the state transition diagram (Fig. 6.1). Now construct the matrix of rates of transition (from row to column)

$$Q = \begin{pmatrix} -\lambda & \lambda & 0 \\ \mu & -\mu-\lambda & \lambda \\ 0 & \mu & -\mu \end{pmatrix}.$$

Next write the equations arising from

$$Q^{\mathrm{Tr}}\boldsymbol{P} = (0),\ \Sigma p_i = 1,$$

where p_i is the probability of S_i and $\boldsymbol{P}$ the vector of p_is

$$-\lambda p_1 + \mu p_2 + 0p_3 = 0$$
$$\lambda p_1 - (\mu + \lambda)p_2 + \mu p_3 = 0$$
$$0p_1 + \lambda p_2 - \mu p_3 = 0$$
$$p_1 + p_2 + p_3 = 1.$$

Note that four equations are needed because the sum of any two of the first three is equal to the other

$$p_2/p_1 = \lambda/\mu = 1/\rho$$
$$p_3/p_2 = \lambda/\mu = 1/\rho,$$

so

$$p_3/p_1 = 1/\rho^2$$

Divide the last equation by p_1

$$1 + p_2/p_1 + p_3/p_1 = 1/p_1$$
$$1 + 1/\rho + 1/\rho^2 = 1/p_1$$
$$A = p_1 + p_2 = (1 + 1/\rho)/(1 + 1/\rho + 1/\rho^2)$$
$$\rho = a/(1 - a),\ \text{by definition}$$
$$A = a/(1 - a + a^2)\ \text{QED}.$$

2. Let $n = 2$, $m = 1$, $r = 2$, S = switch success probability. In this case we have:

   S_1 one running, other standby;
   S_2 one running, one under repair;
   S_3 both under repair *or* one and switch under repair with the second inoperable because of switch failure.

   From S_1 the rate to S_2 is λS (switch works), and to S_3 it is $\lambda(1 - S)$ (switch fails). The switch is not involved in recovery from S_2 to S_1 so that rate is μ, but it may fail (again) when recovering from system failure.

   The state-space diagram for this system is depicted in Fig. 6.2.

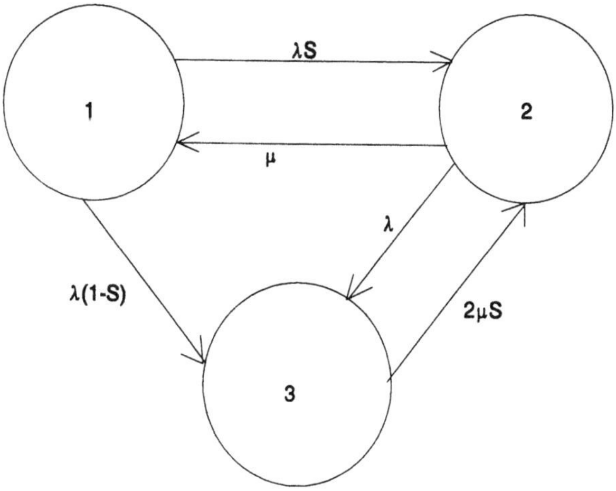

Fig. 6.2 State-space diagram for Problem 6.2.

$$Q = \begin{pmatrix} -\lambda & \lambda(1-S) & \lambda S \\ \mu & -\mu-\lambda & \lambda \\ 0 & 2\mu S & -2\mu S \end{pmatrix}.$$

This matrix assumes that the repair rate for the switch is also μ. Otherwise an answer in terms of *a* and *S* only is not possible.

$$-\lambda p_1 + \mu p_2 + 0p_3 = 0$$
$$\lambda(1 - S)p_1 - (\mu + \lambda)p_2 + 2\mu S p_3 = 0$$
$$\lambda S p_1 + \lambda p_2 - 2\mu S p_3 = 0$$
$$p_1 + p_2 + p_3 = 1$$
$$p_2/p_1 = \lambda/\mu = 1/\rho.$$

Substitute $\lambda p_1 = \mu p_2$ in third equation, giving

$$(\mu S + \lambda)p_2 - 2\mu S p_3 = 0$$
$$p_3/p_2 = (\mu S + \lambda)/2\mu S$$
$$p_3/p_1 = (\lambda\mu S + \lambda^2)/2\mu^2 S = 1/2\rho + 1/2\rho^2 S$$
$$A = (1 + 1/\rho)/(1 + 1/\rho + 1/2\rho + 1/2\rho^2 S).$$

Substituting $\rho = a/(1 - a)$ and omitting some algebra

$$A = 2aS/(1 - a + aS + a^2 S), \quad \text{QED.}$$

PROBLEM 6.2

A productive system consists of three stages, as in Table 6.1. Find the expected long term output rate of the system and the overall MTBF and MTTR.

Table 6.1 Data for Problem 6.2

Stage	*No. fitted*	*No. to run for normal operation*	*Relative max. output per item %*	*MTBF (h)*	*MTTR (h)*
1	3	3	40	2000	4
2	2	1	100	3500	6
3	1	1	120	5000	2

ANSWER

First find the unit availabilities

$$a_1 = 2000/2004 = 0.998\,004 \quad \text{(active parallel)}$$

$$a_2 = 3500/3506 = 0.998\,289 \quad \text{(standby)}$$

$$a_3 = A_3 = 5000/5002 = 0.999\,600$$

$$A_2 = 2a_2/(1 + a_2^2) = 0.999\,998\,5 \quad \text{assuming unlimited repairs.}$$

The combined availability of stages 2 and 3 is, by the product approximation,

$$A_{2,3} = 0.995\,998\,5 \quad \text{with output 100.}$$

The output will be restricted to 100 if all three of stage 1 items operate, to 80 for two, and 40 for one. Total expected output is therefore

$$E_{sys} = A_{2,3}[100a_1^3 + 80 \times 3 \times a_1^2(1 - a_1) + 40 \times 3 \times a_1(1 - a_1)^2]$$
$$= 99.48\%.$$

MTBF and MTTR are imprecise terms for productive systems. We will assume that what is required is the mean time between restrictions in output rate. In stage 1 output is restricted by any failure. The stage 'MTBF' is therefore 2000/3 assuming that the distributions are exponential. Double failures will be relatively rare. In stage 2 we must first find p_2 since all transitions to the failed state occur from S_2

$$Q = \begin{pmatrix} -\lambda & \lambda & 0 \\ \mu & -\mu-\lambda & \lambda \\ 0 & 2\mu & -2\mu \end{pmatrix}$$

$$-\lambda p_1 + \mu p_2 + 0p_3 = 0$$

$$\lambda p_1 = (\mu + \lambda)p_2 + 2\mu p_3 = 0$$

$$0p_1 + \lambda p_2 - 2\mu p_3 = 0$$

$$p_1 + p_2 + p_3 = 1$$

$p_2/p_1 = 1/\rho$, where $\rho = \mu/\lambda = \text{MTBF/MTTR} = 3500/6$

$p_3/p_2 = 1/2\rho$, therefore $p_3/p_1 = 1/2\rho^2$

$$\begin{aligned} p_2 &= (1/\rho)/(1 + 1/\rho + 1/2\rho^2) \\ &= (6/3500)/(1 + 6/3500 + 36/2 \times 3500^2) \\ &= 0.0017113. \end{aligned}$$

The overall failure rate of the stage is then $p_2\lambda$ and its reciprocal is the stage MTBF, i.e.

$$\theta = 3500/p_2 = 2.045 \times 10^6\,\text{h}$$

which will not be important in the final result.

$$\begin{aligned} \theta_3 &= 5000\,\text{h} \\ 1/\theta_{\text{sys}} &= \sum_i 1/\theta_i = 3/2000 + 1/2.045 \times 10^6 + 1/5000 \\ \theta_{\text{sys}} &= 588.1\,\text{h or 14 to 15 times per year if run continuously} \\ \phi_{\text{sys}} &= \theta_{\text{sys}} \sum_i \phi/\theta_i \end{aligned}$$

for a series system. That is, each ϕ_i is weighted according to the relative frequency of failure of the stage which is $\theta_{\text{sys}}/\theta_i$

$$\begin{aligned} \phi_{\text{sys}} &= 588.1 \times [4 \times 3/2000 + 6/2 \times 2.045 \times 10^6 + 2/5000] \\ &= 3.765\,\text{h}. \end{aligned}$$

This analysis omits double failures in stage 1 from consideration but this is unlikely to make much difference to the estimates. An alternative calculation would be to find the mean time between repair actions to all items and the mean time to repair them as measures of system breakdown maintenance load. In this case the combined MTBF is approximately given by

$$\begin{aligned} 1/\theta_{\text{sys}} &= 3/2000 + 1/3500 + 1/5000 \\ \theta'_{\text{sys}} &= 503.6\,\text{h or 17 to 18 times per year.} \end{aligned}$$

Similarly, ϕ_{sys} would be

$$\begin{aligned} \phi_{\text{sys}} &= 503.6[4 \times 3/2000 + 6/3500 + 2/5000] \\ &= 4.086\,\text{h}. \end{aligned}$$

PROBLEM 6.3

A shipping company seeks to discover whether it would be better to train a technician on board to look after a ship's air-conditioning plant or to accept rare failures to the system and repair failures to redundant items in harbour.

The system consists of three sets of rotating machinery each with MTBF 2000 h, MTTR 16 h and a chilled water distribution system driven by one of two pumps each having MTBF 4500 h, MTTR 12 h. In temperate waters, where the ship spends 45% of its time, one set is usually sufficient. In the tropics at least two are required. The system also cools vital navigation and radar equipment, the loss of which is regarded as very serious. At least one set is required at all times to prevent this event.

1. Find the expected proportion of total time for which the system cannot meet its commitments with and without a technician on board given that the ship's regular voyage is between two temperate ports but passes through the tropics and takes 21 days.
2. Find the mean time between complete system failures affecting the vital equipment in each case.

ANSWER

1. In effect we have a one or two-out-of-three standby system in series with a one-out-of-two standby system

$$a_1 = 2000/2076 = 0.992\,06$$

$$a_2 = 4500/4512 = 0.997\,34.$$

We take the case where a technician is carried first. There is only one so we should really consider the whole system in one big matrix. This is not impossible to do but is really not worthwhile for the extra accuracy in this case, particularly as we have to calculate for two situations. In practice, existing technicians can deal with the pumps.

For temperate waters

$$A_1 = a(a^2 - a + 1)/(2a^2 - 2a + 1)$$
$$= 0.9_65 \text{ (i.e. } 0.999\,999\,5)$$

$$A_2 = 2a/(1 + a^2) = 0.9_565.$$

For tropical waters

$$A_1 = (2a^2 - a^3)/(4 - 8a + 6a^2 - a^3)$$
$$= 0.9_3746$$

A_2 is unchanged.
The product rule is good enough here

$$A_{sys} = 0.45(0.9_65 \times 0.9_665) + 0.55(0.9_3746 \times 0.9_565)$$
$$= 0.9_3857.$$

$1 - A_{sys} = 0.000\,143$ is therefore the downtime proportion with a technician. If there are no technicians except at 21-day intervals then the effective item MTTRs will be increased by $10\frac{1}{2}$ days = 252 h since the average failure will occur halfway between ports. Note that it is not correct to regard the items as 'unmaintained stages' because item failures would be repaired as well as system failures at the next port. We now have

$$a_1 = 2000/2268 = 0.881\,83$$

$$a_2 = 4500/4764 = 0.944\,58.$$

Presumably, in port, repair staff will be unlimited so

$$A_1 = 3a^2(2 - a)(a^3 + 2)$$
$$(n = 3,\ m = 2,\ r = 3 \text{ continue on stage failure})$$
$$= 0.826\,59 \text{ (tropical)}$$

$$A_1 = 3a(1 + a^2)/(2a^3 + 3a + 1)$$
$$= 0.937\,36 \text{ (temperate)}$$

$$A_2 = 2a/(1 + a^2)$$
$$= 0.998\,38$$

$$A_{sys} = 0.998\,38\ (0.45 \times 0.937\,36 + 0.55 \times 0.826\,59)$$
$$= 0.8750$$

$$1 - A_{sys} = 0.1250, \text{ a difference of three orders of magnitude.}$$

Note that it would not make a great deal of difference if the pumps were omitted from the calculation and that most of the problems will arise in the tropics in mid-voyage with the compressor sets. 12% downtime is probably unacceptable so the technician would have to be trained.

2. An exact answer to this part can be obtained by matrix algebra, but a shortcut is available to give an approximate answer which is probably good enough to indicate the order of magnitude and make the decision. The availability of the system, against total failure is the temperate water figure, i.e.

$$A_{sys} = 0.9_65 \times 0.9_565 = 0.9_56 \text{ with a technician}$$

$$1 - A_{sys} = 0.0_54$$

or

$$A_{sys} = 0.937\,36 \times 0.998\,38 = 0.935\,84$$

$$1 - A_{sys} = 0.064\,16 \text{ with no technician.}$$

Now ϕ_{sys} is between 12 and 16 hours and will be nearer 16 than 12, say 15 h or say 267 h with no technician. Then

$$\theta_{sys} \approx 15 \times 0.9_56/0.0_54 = 3.75 \times 10^6 \text{ h with technician}$$

$$\theta_{sys} \approx 267 \times 0.935\,84/0.064\,16 = 3895\,\text{h with no technician.}$$

With no technician we expect a complete breakdown once every $3895/(21 \times 24) = 7.73$ voyages.

COMMENT

This problem is in reality even more complex because some failures will not be repairable at sea even if a technician is carried.

REFERENCE

Sherwin, D.J. (1984) *International Journal of Production Research*, **22**(1), 109–20.

CHAPTER 7

Systems of multistate items

We have identified several general cases in which the usual two-state (up or down, working or failed) assumption about items is not valid. The first of these, open- and short-circuit failures, is not confined to electrical systems but can also be found in fluid systems where a valve failed shut may be considered analogous to an open-circuit electrical failure and *vice versa*. Then there is the case of the item which suffers some failures which are repairable and others which are not, or which cannot be repaired immediately or during a mission over which the reliability of the system is to be estimated. Finally we consider items with three and more states and develop some general methods for dealing with systems of multi-state items.

Open- and short-circuit reliability can be and here is dealt with by simple probability modelling. In the sections on repairable and non-repairable failures and the final general section on multi-state systems we make use of the matrix techniques developed in Chapter 5.

7.1 OPEN- AND SHORT-CIRCUIT FAILURES

7.1.1 Introduction

It has elsewhere been assumed that item failure produces a condition as if the item were physically removed from the system or circuit. In electrical, electronic, mechanical and fluid systems, failures emulating bypass of the item are also possible. These can be termed open- and short-circuit failures, respectively. In this section only, the terms series and parallel refer to physical arrangement rather than reliability block diagram conventions.

This section is based upon a paper by Jenney and Sherwin (1986) and subsequent correspondence. We are grateful to the editor and Professor Jenney for permission to republish this work in revised form.

The paper was about reliability and risk analysis of systems of identical, independent items in terms of their probabilities of failure in open and short circuit. The important assumptions were:

1. the system is operational while at least one (or as a special case, more) operational item is connected physically to the input and output of the

system either directly or through other good or shorted items and there is no overall system short circuit.
2. Two or more items so connected, whether in physical series or parallel or any combination have the same effect as a single item, and such redundancy does not alter the item failure probabilities.
3. The probabilities may be those arising from identical, independent distributions over a fixed period with fixed starting conditions or alternatively probabilities of failure on switching on.

The work applies therefore to systems of adjustable or self-adjusting items such as amplifiers, rheostats, or fluid control valves, but not, for example, to systems of fixed resistors if the output values of voltage and current are important. Methods were developed for finding the most reliable physical arrangements of two, three, or four items given the item failure probabilities. These display symmetry arising from the equivalence of the reliability block diagrams for items in physical series failing short and items in physical parallel failing open and *vice versa*. The increment in reliability for each additional item becomes rapidly smaller if the most advantageous arrangement is used and can be negative otherwise. The results for three and four items are presented in the form of handy graphs for choosing the most reliable arrangement. Equations are given for calculating system reliability from item open and short failure probabilities.

The paper also dealt with series-parallel and parallel-series arrays containing any non-prime number of identical items. If both open and short failures are possible, an array is usually more reliable than the same number of items in parallel or in series. Using the most reliable arrangement, reliability increases with number of items, but if the number of items in either direction in an array of either type is fixed, then the number in the other direction has an optimum value, unless either of the item probabilities is zero.

Systems can fail either open or short circuit. The paper also considered the case where one mode (open or short) was less desirable than the other because of differences in cost or (dangerous) consequences. This allowed the work to be applied to risk analysis. Much of the theory has been previously published fragmentarily or has been known in some circles but not published. The paper gathered these fragments together in coherent form, filled in some of the gaps, and provided limited but useful extensions.

7.1.2 Notation and special terminology

q,s	Unconditional probability an item fails open, short, $q + s < 1$.
V	Item reliability, $V = 1 - q - s$, $0 < V < 1$.
P	Proportion of item failures that are open circuit $P \equiv q/(q + s)$, $0 < P < 1$.

u	Undesirability factor in risk calculation.
Z	Risk (defined below).
m,n	Number of rows, columns in an array, also used as subscripts.
array	A rectangular arrangement of a non-prime number of items which can be series-parallel or parallel-series.
series-parallel array	An array consisting of a number of parallel arrangements of equal numbers of items arranged in series.
parallel-series array	An array consisting of a number of series arrangements of equal numbers of items arranged in parallel.
open, open-circuit	Failure of an item has the same effect upon the system as removal of the item, the free ends of connectors being blanked off to prevent loss of fluid etc. as appropriate.
short, short-circuit	Failure of an item has the same effect upon the system as replacement of the item with a simple physical connection, e.g. a wire or plain pipe as appropriate to the case.

The terminology for open or short failures is electrical. A fluid valve failed shut is an open failure and *vice versa*. Open and short failures are mutually exclusive.

(a) q − s *symmetry*

Two arrangements of the same number of items exhibit $q - s$ symmetry if their reliability expressions are such that interchange of q and s in the formulations converts one to the other. It follows that the reliability block diagram of one against short failure is the same as the reliability block diagram of the other against open failure and *vice versa*.

(b) *Arrangement*

$(z, xs + yp/2s)$ consists of z items arranged as x items in series in series with $(+)y$ items in parallel with each other and also with $(/)$ a sub-arrangement of two items in series. $(z = x + y + 2)$. This notation is also used as a subscript to R to indicate the reliability of such an arrangement. For $x = y = 2$, $z = 6$ the arrangement is shown in Fig. 7.1 Note $(6, 2s + 2s/2p) \equiv (6, 2s/2p + 2s)$.

(c) *Assumptions*

1. All the items in the system are identical and independent. Where their open and short failure probabilities are determined by failure time

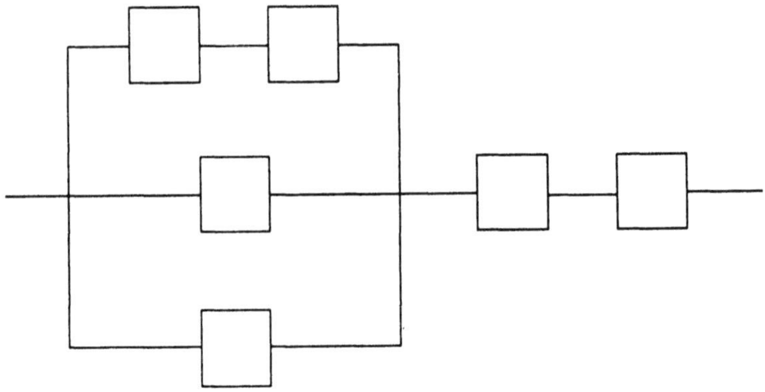

Fig. 7.1 Arrangement $(6, 2s/2p+2s)$ to illustrate method of defining arrangements.

distributions the items are independent and identically distributed.

2. Except where more than one operational path is required, any one item can perform the function of the system but the output is unaffected by passing through additional good items or by being split between several good paths. In exceptional cases a minimum of j separate parallel paths each containing a minimum of k good items is required; other conditions are unaltered.
3. Items exist in one of three states, namely, good, failed open and failed short. These states are mutually exclusive and exhaustive. As a result of this assumption, systems also have only two classes of failure, open and short.
4. The failure probabilities are unaffected by the number of parallel paths among which the function is divided or by the failure of other items.

These assumptions are often implicit in conventional reliability theory, and often correct for electrical/electronic systems. For mechanical and fluid systems, they need to be re-stated with a warning to avoid misapplication. For example, real fluid valve failure modes include 'excessive leakage to atmosphere' and 'stuck part-open', which do not conform to the rules and assumptions above. Such non-conformity may affect the applicability of the theory.

7.1.3 Series and parallel systems subject to open and short failure

(a) One path required

For series systems failure occurs when all n items fail short with probability s^n or when the first item fails open; probability $1 - (1 - q)^n$. The reliability is therefore

$$R_{ns} = (1 - q)^n - s^n, \; n = 1, 2, 3, 4, \ldots. \tag{7.1}$$

For parallel systems to fail, either all m items must fail open or any item short

$$R_{m\mathrm{p}} = (1 - s)^m - q^m, \; m = 1, 2, 3, 4, \ldots . \tag{7.2}$$

Note the $q - s$ symmetry between (7.1) and (7.2).

(b) j or more parallel or series items required

For parallel systems, failure occurs if more than $m - j$ items fail open or if any item fails short

$$R_{m\mathrm{p}j} = (1 - s)^m - \sum_{i=0}^{j-1}\left[\binom{m}{i}q^{m-i}(1 - q)^i\right]. \tag{7.3}$$

For series systems there is only one path but an analogous situation arises if short failure of more than $n - j$ items out of n is regarded as a system failure. In this case, for n items in series

$$R_{n\mathrm{s}j} = (1 - q)^n - \sum_{i=0}^{j-1}\left[\binom{n}{i}s^{n-i}(1 - s)^i\right]. \tag{7.4}$$

Note the $q - s$ symmetry between (7.3) and (7.4).

EXAMPLE 7.1

An electronic circuit consists, in part, of six identical amplifiers in (electrical) parallel, two of which are redundant. Each amplifier has a total constant hazard rate of 1×10^{-5} per hour and it is estimated that 40% of failures are open circuit. It is required to find the combined amplifier reliability over 10 000 hours.

SOLUTION

$$\begin{aligned} s &= 1 - \mathrm{e}^{-6\times 10^{-2}} = 0.058\,235 \\ q &= 1 - \mathrm{e}^{-4\times 10^{-2}} = 0.039\,211. \end{aligned} \tag{7.5}$$

Then apply equation (7.3), putting $m = 6$, $i = 4$ and q and s as above to obtain $R = 0.696\,58$.

7.1.4 Optimum number in parallel or series

It is demonstrated below that where both open and short failures are possible, simple series and parallel arrangements are seldom the most reliable arrangement of a given number of items. However, if one of these arrangements is specified there is an optimum number of items m^* or n^*. In a parallel arrangement, the reliability against open system failures increases with number of items whilst that against short system failure falls and *vice versa* for series systems. Von Alven (1965) and Shooman (1968) both give the following method for finding m^* and n^*, when $j = 1$.

Recall that for any function $f(x)$ of x

$$df(x)^n/dn = f(x)^n \ln f(x). \tag{7.6}$$

Differentiation of (7.1) and (7.2) with respect to n and m and equating to zero results in

$$n^* = \ln[\ln(s)/\ln(1-q)]/\ln[(1-q)/s] \tag{7.7}$$

$$m^* = \ln[\ln(q)/\ln(1-s)]/\ln[(1-s)/q)]. \tag{7.8}$$

The values of m^* and n^* must of course be integers so it is necessary to try the integer values either side of the calculated value. Note the q-s symmetry between (7.7) and (7.8). If $q = s$, $m^* = n^* = 1$ or 2, and (7.1) and (7.2) both give $R = 1 - 2q$ which is equal to V, the reliability of a

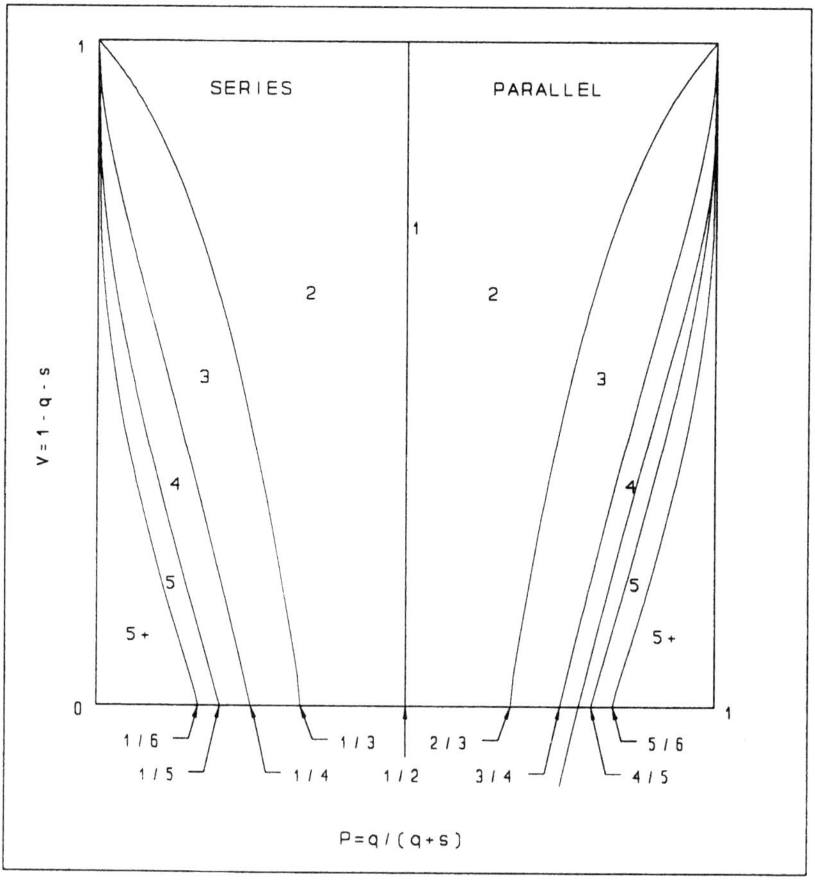

Fig. 7.2 Zones in which two, three, four, five and more than five items in parallel or series have higher reliability than any other simple series or parallel arrangement.

single unit. The changeover from series to parallel for maximum reliability occurs when

$$P = q/(q + s) = 1/2 \text{ for all } q \text{ and } s \text{ for which } q + s < 1.$$

At low values of P, additional items in series give higher R than fewer. Similarly at high values of P more items in parallel raise R. As V approaches 1 there is less and less advantage in more units and P must be higher or lower to make an additional unit worthwhile. In the other limit as V tends to zero the optimal changeover from n to $n + 1$ items in series occurs at $P = 1/(n + 1)$ and in parallel at $P = m/(m + 1)$. Figure 7.2 gives the zones of optimality for simple parallel and series systems.

7.1.5 Arrays

(a) Introduction

An array is a rectangular arrangement of a non-prime number of items. We examine both types of array, parallel-series (ps) and series-parallel (sp) (Figs 7.3 and 7.4). In a ps-array there are m parallel rows each

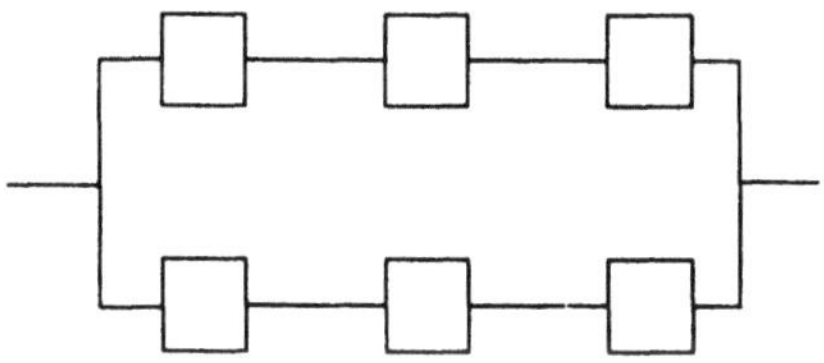

Fig. 7.3 2×3 ps array.

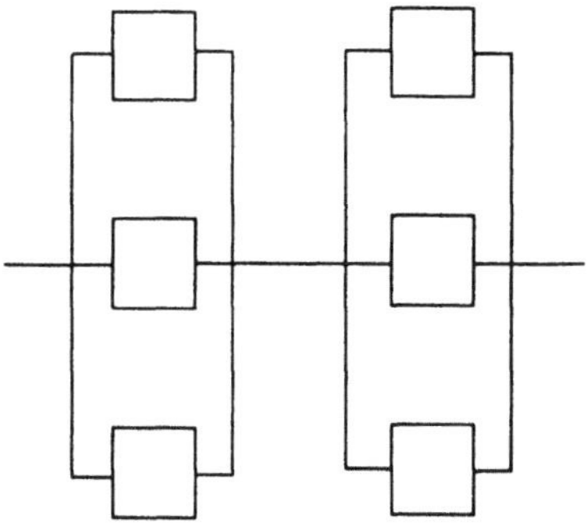

Fig. 7.4 3×2 sp array.

consisting of n items in series. In an sp-array there are n arrangements each with m items in parallel, arranged in series. In each case the convention adopted is that the number of rows is given before the number of columns.

(b) *Parallel-series array – single path required*

In this case, failure occurs if there is an open failure in each of the paths, or if all the n items in a path fail short-circuit. The system fails either open or short, it cannot fail by a combination of the two modes. The probability of not failing in one mode is included in the probability of failing in the other. Just as for a single item $V = 1 - q - s$, it is equally true than $V_{sys} = 1 - Q - S$, where Q is the probability of system failure in open circuit and S in short circuit. In this case

$$Q_{ps} = [1 - (1 - q)^n]^m$$
$$S_{ps} = 1 - (1 - s^n)^m$$

and the reliability is

$$R_{ps} = 1 - Q_{ps} - S_{ps} = (1 - s^n)^m - [1 - (1 - q)^n]^m. \qquad (7.9)$$

(c) *Series-parallel array – single path required*

In this case, failure occurs if there is a short failure in each of the simple parallel sub-arrays, or if all the items in any of these sub-arrays fail open.

$$Q_{sp} = 1 - (1 - q^m)^n$$
$$S_{sp} = [1 - (1 - s)^m]^n$$

and the reliability is

$$R_{sp} = 1 - Q_{sp} - S_{sp} = (1 - q^m)^n - [1 - (1 - s)^m]^n. \qquad (7.10)$$

Note the $q - s$ symmetry, in which m and n are also transposed, between (7.9) and (7.10).

(d) *Optimal arrays with* m *or* n *fixed*

For a ps-array, n fixed, and only one path required, the optimal number of rows m^* can be found by using equation (7.1) to combine the items in a row into one equivalent item in which $Q = 1 - (1 - q)^n$ and $S = s^n$. We can then substitute Q and S for q and s in equation (7.8) to find m^*. Similarly for a single path in an sp-array and a fixed value of m it is possible to optimise n from equation (7.7) using $Q = q^m$ and $S = 1 - (1 - s)^m$. In both cases it is assumed that only one good item is required, i.e. $k = 1$ in the terminology used above. Optimization with $k > 1$ is left as an exercise for the reader.

7.1.6 Non-array arrangements

In this section arrangements of 3 and 4 items which are not arrays are examined. Arrangements of five items were also examined in the research but we did not consider them worth including because the differences in reliability between them and the best four-item arrangements are negligibly small for practical values of q and s. With more than four items it is necessary to devise reliable systematic means of listing all the possible arrangements to be certain that every one has been considered. Some severely lopsided arrangements are never optimal over all other arrangements of the same number of items. For non-prime total numbers, arrays tend to be optimal over other arrangements, where $V = 1 - q - s$ is high and q and s were of the same order of magnitude.

(c) Non-array three-item arrangements

There are two arrangements of three items other than parallel and series which must be considered when both open and short failures are possible. These are (3, 1 + 2p) and (3, 2s/1), i.e. one item in series with two in parallel, and one item in parallel with two in series (Fig. 7.5).

Arrangement (3, 1 + 2p)

Pr{open failure}: $q + q^2 - q^3$
Pr{short failure}: $2s^2 - s^3$
Reliability: $R = 1 - q - q^2 + q^3 - 2s^2 + s^3$. (7.11)

Arrangement (3, 2s/1)
This is q-s-symmetrical with (3, 1 + 2p)
Reliability: $R = 1 - s - s^2 + s^3 - 2q^2 + q^3$. (7.12)

(b) Non-array four-item arrangements

There are six other arrangements of four items besides series, parallel, 2 × 2 sp and 2 × 2 ps arrays.

Arrangement (4, 2s/2p)

Pr{open failure}: $(2q - q^2)q^2$
Pr{short failure}: $2s - s^2(2s - s^2)$
Reliability: $R = 1 - q^2(2q - q^2) - 2s(1 - s^2) - s^4$. (7.13)

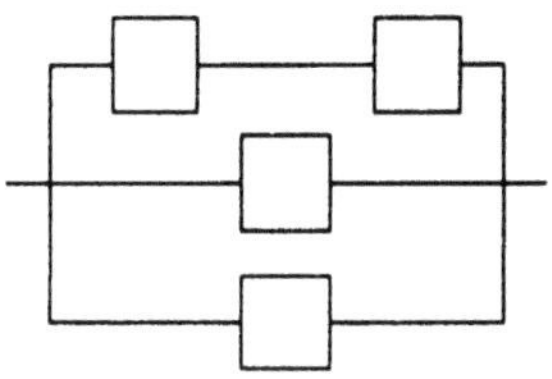

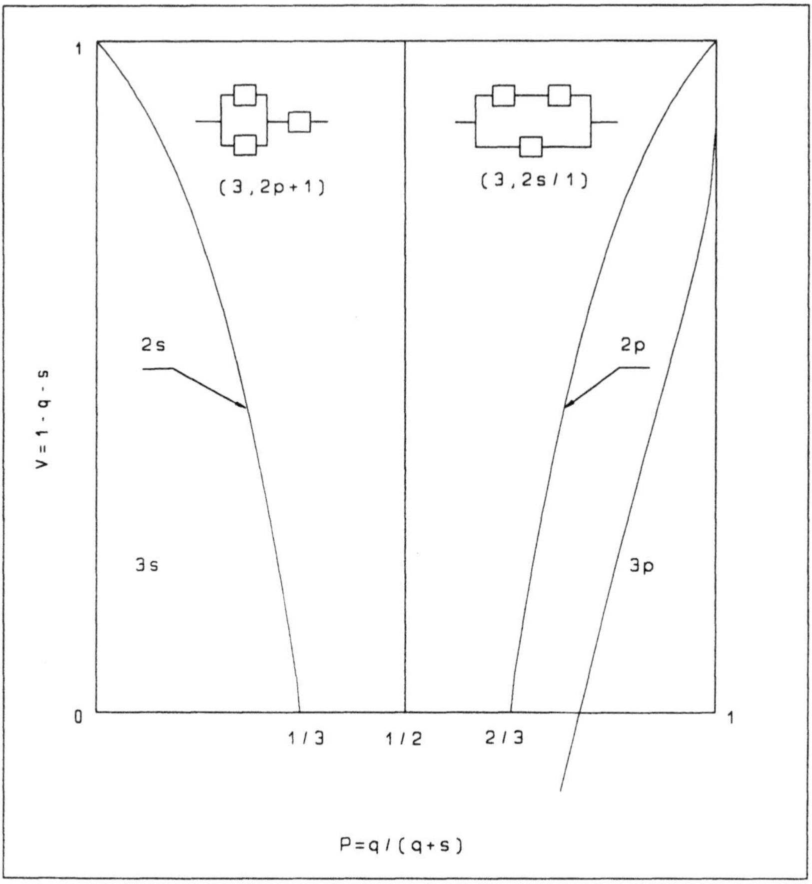

Fig. 7.5 Zones of highest reliability for arrangements of up to three items. On the curved lines two are as reliable as three.

Arrangement (4, 2s + 2p)

Pr{open failure}: $2q - q^2(2q - q^2)$
Pr{short failure}: $s^2(2s - s^2)$
Reliability: $R = 1 - s^2(2s - s^2) - 2q(1 - q^2) - q^4.$ (7.14)

Note the q-s-symmetry between (4, 2s/2p) and (4, 2s + 2p).

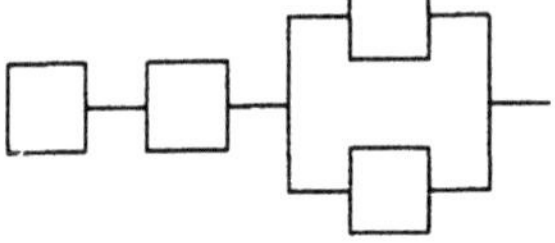

Arrangement (4, 3s/1)

Pr{open failure}: $q - q(1-q)^3$
Pr{short failure}: $s + s^3 - s^4$
Reliability: $R = 1 - q + q(1-q)^3 - s - s^3 + s^4$. (7.15)

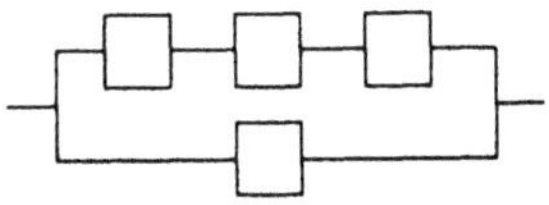

Arrangement (4, 3p + 1)

Pr{open failure}: $q + q^3 - q^4$
Pr{short failure}: $s - s(1-s)^3$
Reliability: $R = 1 - s + s(1-s)^3 - q - q^3 + q^4$. (7.16)

Note the q-s-symmetry between (4, 3s/1) and (4, 3p + 1).

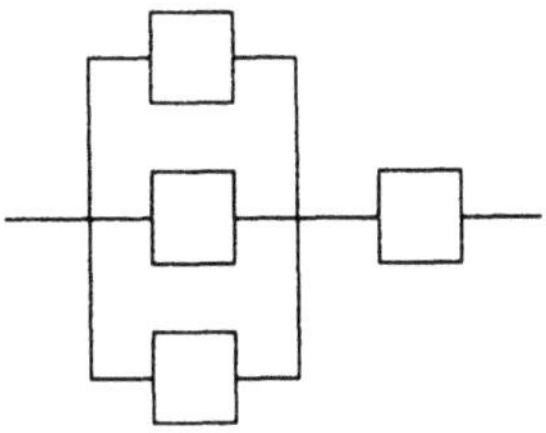

Arrangement (4, 2s/1 + 1)

Pr{open failure}: $q + 2q^2 - 3q^3 + q^4$
Pr{short failure}: $s^2 + s^3 - s^4$
Reliability: $R = 1 - q - 2q^2 + 3q^3 - q^4 - s^2 - s^3 + s^4$. (7.17)

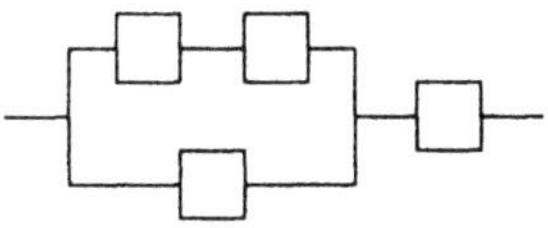

Arrangement (4, 1/(1 + 2p)

Pr{open failure}: $q^2 + q^3 - q^4$
Pr{short failure}: $s + 2s^2 - 3s^3 + s^4$
Reliability: $R = 1 - s - 2s^2 + 3s^3 - s^4 - q^2 - q^3 + q^4$. (7.18)

Note the q-s-symmetry between (4, 2s/1 + 1) and (4, 1/(1 + 2p)).

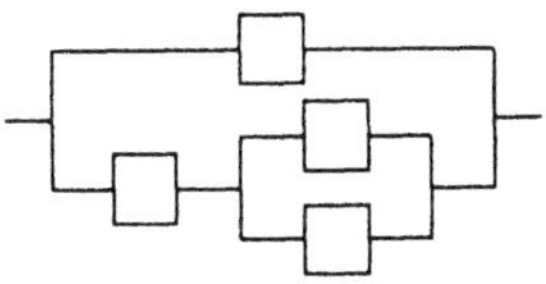

(b) Handy graphs for three and four items

Figures 7.5 and 7.6 are *P v. V* diagrams for three- and four-item arrangements, respectively. The diagrams are for a single path and one good item per path. The lines divide regions in which the specified arrangements give the highest reliability. These diagrams, drawn from computer output are not strictly to scale. If *P* and *V* meet near a line, two or three calculations should be made using the equations above to discover the most reliable arrangement. Three calculations may be required because on some of the lines two items are as good as three or three as good as four, as indicated. This is discussed below.

(c) Single and one-out-of-two systems

It is instructive to calculate the conditions where $R_{2s} = R_{2p}$ in terms of *P* and *V*. From the definitions

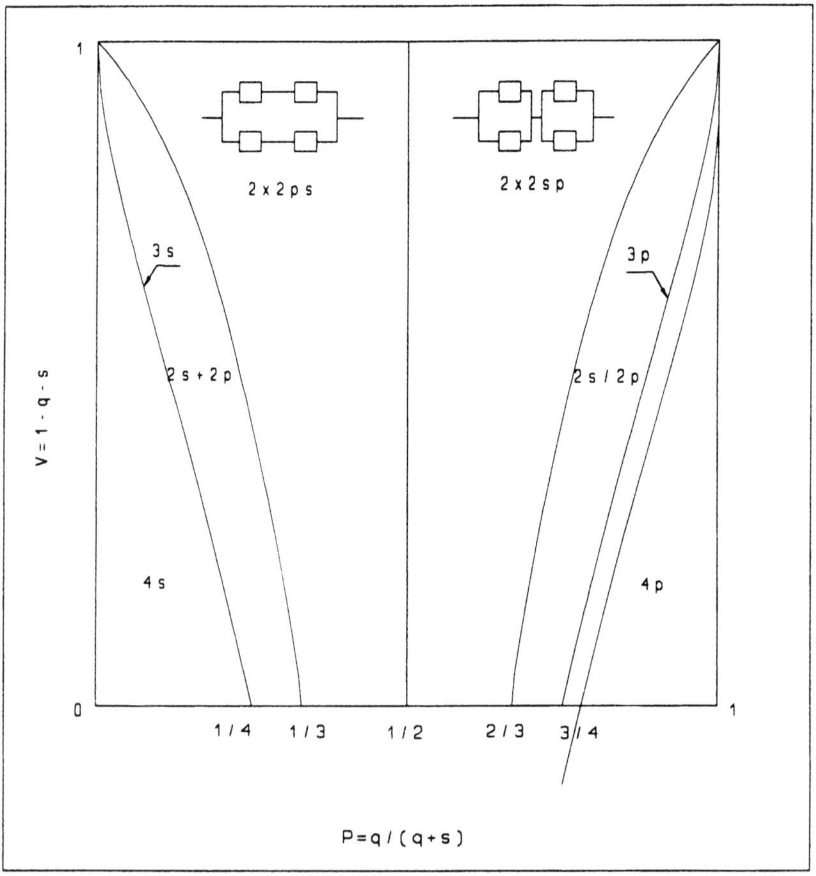

Fig. 7.6 Zones of highest reliability for arrangements of up to four items. Note where three items are as reliable as four.

$$q = P(1 - V), s = (1 - P)(1 - V). \qquad (7.19)$$

One solution is obviously $q = s$, in which case $P = 1/2$ and V can take any value. $V = 0$ and $V = 1$ are the other solutions. When $R_{2s} = R_{2p}$ both are also equal to V, the single unit reliability. With this result in mind, we investigated boundary conditions in systems with three and four items.

(d) Systems of up to three items

A computer was used to find the most reliable arrangement of up to three items for 40000 pairings of P and V, i.e. increments of 0.005 from 0 to 1. The results suggested (Fig. 7.5) that the lines on the P v. V diagram at which $R_{3s} = R_{3,1+2p} = R_{2s}$ and $R_{3p} = R_{3,2s/1} = R_{2p}$ start at the top corners of the diagram and curve down to 1/3 and 2/3 on the P-axis. This was proved algebraically, the equations of the lines being, respectively,

$$V = (P - 1)(3P - 1)/(3P^2 - 3P + 1) \qquad (7.20)$$

$$V = P(3P - 2)/(3P^2 - 3P + 1). \qquad (7.21)$$

By lowering the number of significant figures in the computerized comparative calculations it was shown that it was not often worthwhile having more than three units unless P was close to either 0 or 1 (nearly all failures of one mode) and/or V was low (unlikely). These algorithms are used more for prediction than autopsy, so the precision with which q and s and thus P and V are known is likely to be low which also argues for simpler one-, two- or three-element systems, at least until tests have produced some failure data. When V is high other arrangements are more likely to be optimal than series or parallel. It is not the case for $P = 0.5$, and a limit of three items, that a single element or a two-element system is equally reliable (except at $V = 0,1$).

(e) Systems up to four elements

The areas of the P v. V diagram (Fig. 7.6) over which the non-array arrangements (4, 2s + 2p) and (4, 2s/2p) are optimal are small and the advantage over 3s or 3p marginal. If Fig. 7.5 is modified to allow 4s and 4p, then 3s, 3p appear in the same areas as (4, 2s + 2p), (4, 2s/2p) on Fig. 7.6. The most common arrangements in service are probably 2 × 2 sp and 2 × 2 ps arrays. This is certainly true of fluid valve systems. It is important that they are most reliable over well over half of Fig. 7.6. (4, 3s/1) and (4, 3p + 1) are never optimal. This is because the lone item is weak where the remainder of the system is strong, against short system failure for (4, 3s/1) or open system failure for (4, 3p + 1). It was shown that (4, 2s/1 + 1) and (4, 1/(1 + 2p)) were also never optimal.

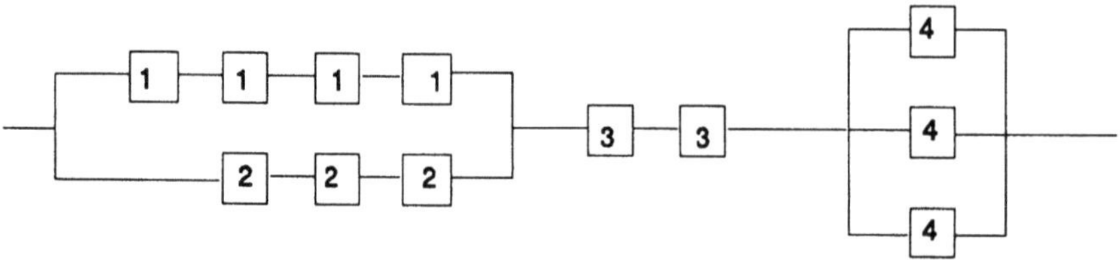

Fig. 7.7 Arrangement of 12 items.

(f) Systems of more than four items and more than one kind

The reliability of non-array arrangements of more than four items and of more than one kind can be calculated by breaking the system into subsystems of two, three or four items or arrays whose reliability can be found and then applying the general rules of parallel and series arrangements. For example, take the arrangement shown in Fig. 7.7. Assign suffices 1, 2, 3, and 4 to the four-in-series top subsystem, the three-in-series subsystem in parallel with it, the two-in-series subsystem in series with them, and the three-in-parallel subsystem on the right, respectively, as annotated in the figure

$$\begin{aligned} Q_1 &= 1 - (1 - q_1)^4, & S_1 &= s_1^{\,4} \\ Q_2 &= 1 - (1 - q_2)^3, & S_2 &= s_2^{\,3} \\ Q_3 &= 1 - (1 - q_3)^2, & S_3 &= s_3^{\,2} \\ Q_4 &= q_4^{\,3}, & S_4 &= 1 - (1 - S_4)^3. \end{aligned}$$

The system is now reduced to Fig. 7.8.

Now if all the Q_i and S_i were equal, we could apply equation (7.14), but they are not, even if all the q_i and s_i are the same throughout. However, we can still apply the logic which led to equation (7.14) as follows. Failure will occur in open circuit if either both 1 and 2 fail open or 3 or 4 singly. The probability of this event is

$$Q_{\text{sys}} = Q_3 + Q_4 + Q_1Q_2 - Q_3Q_4 - Q_1Q_2(Q_3 + Q_4) + Q_1Q_2Q_3Q_4.$$

Similarly,

$$S_{\text{sys}} = S_3S_4(S_1 + S_2) - S_1S_2S_3S_4$$

and finally

$$R_{\text{sys}} = 1 - Q_{\text{sys}} - S_{\text{sys}}. \tag{7.22}$$

Note that this system could still have been solved even if all 12 items in it had been different.

If Fig. 7.8 represents a series system with redundancy on the first stage (1 and 2) in which all three stages must operate then the problem is different. A short in any block is now a system failure and either block 1 or block 2, plus block 3 and block 4 must be free of open failures.

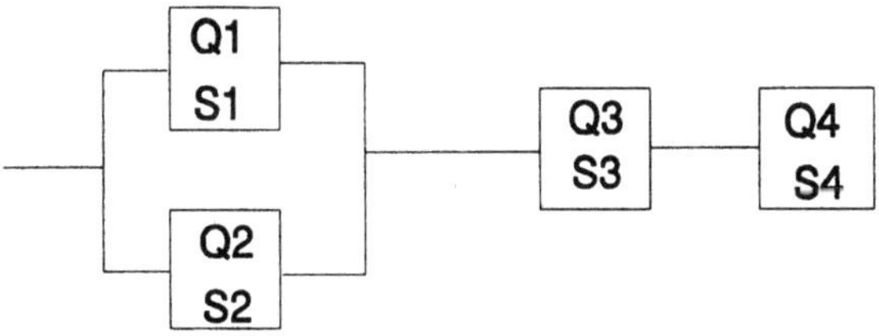

Fig. 7.8 Reduced version of the system in Fig. 7.7.

For the first stage, blocks 1 and 2

$$R_{12} = 1 - S_1 - S_2 + S_1S_2 - Q_1Q_2.$$

For blocks 3 and 4

$$R_{34} = (1 - Q_3 - S_3)(1 - Q_4 - S_4).$$

And for the whole system

$$R_{sys} = R_{12} \times R_{34}.$$

Note that Fig. 7.8 is a true reliability block diagram as well as a representation of the physical arrangement of the blocks.

7.1.7 Standby redundancy with open and short failures

In a simple one-out-of-*n* standby system it makes no difference to the reliability whether the failure was short or open. However, the switch used to change over may have four modes of failure which were examined in Chapter 4. Two of these modes equate to open and short failure so that the one-out-of-two system becomes, if the fluttering and spurious switching modes are ignored, a three-element system.

EXAMPLE 7.2 (Fig. 7.9)

The detonating system for a bomb consists of an altitude-activated arming switch which in turn activates an automatic changeover switch which then

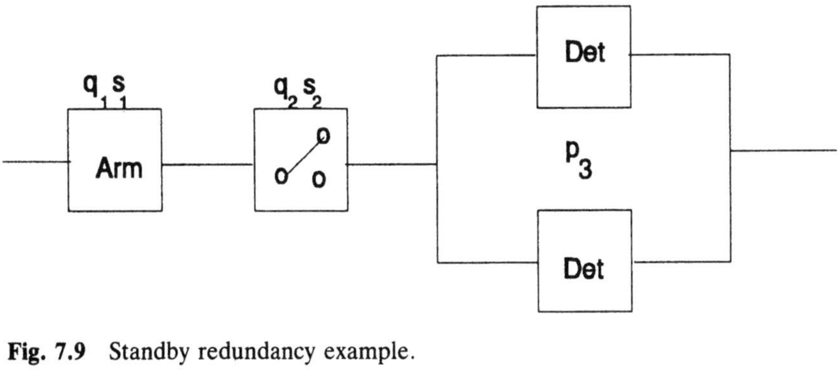

Fig. 7.9 Standby redundancy example.

tries one detonator and, if necessary, the other. Calculate the probabilities of short and open failure of this system in terms of the individual open and short failure probabilities of the three types of component.

SOLUTION

Use subscripts 1, 2 and 3 for the devices in the operating sequence. The detonators either operate or fail to do so – the concept of open and short failure cannot apply to them. Short failure of both switches would produce premature detonation of the bomb if the detonator worked. The probability of this is $s_1s_2p_3$, where p_3 is the probability a detonator fires when activated. It is p_3 rather than $2p_3 - p_3^2$ because presumably the switch is not activated correctly but shorted and therefore only one detonator would be connected; it would not change over in this condition. This is the probability of premature detonation of the bomb. An open system failure would leave the bomb unexploded when it was desired that it detonate. The probability of this is

$$q_1 + q_2 + (1 - p_3)^2 - q_1q_2 - (q_1 + q_2)(1 - p_3)^2 + q_1q_2(1 - p_3)^2.$$

COMMENT

The individual short failure probabilities would be calculated for the conditions and the time for which the detonators were present in the bomb. It is wise to keep this time as short as possible. The open-circuit probabilities are mainly those of failure to operate or maloperation rather than time related. The short failure probabilities would be very small, but the fact that accidents do sometimes occur with munitions indicates that they are not zero.

7.1.8 Risk analysis

(a) One mode less desirable

'Best' does not always equate with 'most reliable'. Frequently one mode of failure is more costly or dangerous than the other. For example, if open failure must be avoided regardless of the resulting probability of short failure then obviously a parallel system with as many elements as possible is optimal. Where the only data are q and s and the undesirability or cost factor u (such that short system failure is u times less desirable or more costly than open failure), the objective is to minimize the risk by choosing amongst the permitted arrangements. These can be ranked by modifying their unreliability expressions such that every term in s is multiplied by u. If the factor applies to item rather than system failure, as it would if the principal consideration was the cost of individual repairs for the two modes, then terms in s^i should be multiplied by iu. If open failure is less desirable, the factor u can be applied to terms in q instead of s.

EXAMPLE 7.3

Reducing valves for natural gas distribution are fitted in groups of up to four for a single function. They fail open at a rate 3×10^{-6}/h and shut at a rate of 7×10^{-6}/h. The valves are refurbished to good-as-new at system failure and every 2 years from previous failure or refurbishment. Overpressure downstream is regarded as 10 times more serious than cessation of supply because of the safety implications. It is required to examine critically the gas utility's choice of a 2×2 ps array for such valve-groups. (This situation is real but the figures are not.)

SOLUTION

A valve failing open is analogous to short failure and *vice versa*. Assume exponentially distributed failure times.

There are 8760 hours in a 365-day year so

$$q = 1 - \exp(-7 \times 10^{-6} \times 8760 \times 2) = 0.1154$$
$$s = 1 - \exp(-3 \times 10^{-6} \times 8760 \times 2) = 0.0512$$

for the overall interval of two years, giving

$$P = 0.6927,\ V = 0.8334.$$

From Fig. 7.6 the most reliable arrangement is 2×2 sp and the reliability from equation (7.10) is 0.9636.

However, 2×2 sp is not the most effective arrangement against short-circuit failure when $u = 10$. The choice is obviously between 2×2 sp, 4s, 3s, (4, 2s + 2p) and 2×2 ps. The risks, Z are calculated as

$$
\begin{aligned}
Z_{2\times 2\text{ps}} &= 20s^2 - 10s^4 + \{1 - (1 - q)^2\}^2 &&= 0.099\,66 \text{ (least risky)}\\
Z_{2\times 2\text{sp}} &= 1 - (1 - q^2)^2 + 10\{1 - (1 - s)^2\}^2 &&= 0.126\,01\\
Z_{3\text{s}} &= 1 - (1 - q)^3 + 10s^3 &&= 0.309\,13\\
Z_{4\text{s}} &= 1 - (1 - q)^4 + 10s^4 &&= 0.387\,74\\
Z_{4,2\text{s}+2\text{p}} &= 10s^3(2 - s) + 2q(1 - q^2) + q^4 &&= 0.2305.
\end{aligned}
$$

So 2×2 ps is the least risky arrangement. The Z figures are not probabilities but risk or probability-cost comparators. The reliability of 2×2 ps for these values of q and s is 0.9475 which should be compared with 0.9636 for 2×2sp. However, the system probabilities of short and open failures, S_{ps} and Q_{ps}, are 0.005 25 and 0.0473, respectively, nearly an order of magnitude different. Also note that $S_{sp} = 0.009\,96$ which is nearly twice S_{ps}.

7.2 REPAIRABLE AND NON-REPAIRABLE FAILURES

7.2.1 Introduction

In this section we examine the calculation of reliability and availability for systems of items which can fail in two distinct ways one of which is repairable and the other not. Complex items often exhibit this behaviour. In theory, any failure can be repaired at a price, but we have seen elsewhere that such an item's life can be ended when it becomes uneconomic to repair. However, the main application of the methodology to be described below is to the classification of failures according to the facilities that are needed to repair them and the availability of these facilities at the point of failure. Thus a non-repairable failure may in fact be one that requires the item to be sent to a central workshop or returned to its manufacturer for repair. There may be more than two modes distinguished in this way. For example failures to deep-mining machinery may be, alternatively, fixable underground, fixable in a surface workshop, fixable by the manufacturer or not fixable at all.

A common problem in military and particularly naval applications involves the requirements at various phases of a putative mission for various minimum numbers of redundant items to be operational. For example, a ship with four main generators probably runs only two most of the time but runs all four during the action phase of a 'standard phased mission' which is specified by the authorities to provide comparisons between competing designs. In this case the distinction is between failures which can be repaired at sea (R) and those which cannot (NR). In a phased mission it may be possible to repair some failures during the less exciting phases, in time for the action phase. The results of such calculations affect the number of ships or aircraft or tanks required by the authorities to counter a given perceived threat posed by a potential enemy.

As with open and short circuit failures it is not possible to treat each mode (R, NR), as a separate block on a reliability block diagram because of the phenomenon known as *interference*. This simply means in this case that an item which has failed NR cannot subsequently fail R, and that an item which is failed R is not at risk of NR failure until it has been repaired.

As an illustration of the ways in which matrix techniques can be used for this sort of problem, we consider phased missions of systems of identical items, having constant failure and repair rates. However, it should be appreciated that the technique of successive multiplication of matrices can also be used to deal with cases where the hazard and repair rates are not constant. But in that case the lives of all items since last failure in each mode must be known at mission start.

7.2.2 Markov matrix techniques – rates and probabilities

In Chapters 5 and 6 we described and illustrated the methods of using transition-rate matrices for calculations in respect of systems consisting of items having constant failure and repair rates (whether the 'failure rate' was a hazard rate or a ROCOF made no difference to the mathematics). In this methodology we would represent a system of two identical items subject to repairable failures at rate λ_1, non-repairable failures at rate λ_2 and repairs at rate μ and operated in standby with two repair crews available by the matrix of transition rates Q_1

$$Q_1 = \begin{bmatrix} -\lambda_1-\lambda_2 & \lambda_1 & \lambda_2 & 0 & 0 & 0 \\ \mu & -\mu-\lambda_1-\lambda_2 & 0 & \lambda_1 & \lambda_2 & 0 \\ 0 & 0 & -\lambda_1-\lambda_2 & 0 & \lambda_1 & \lambda_2 \\ 0 & 2\mu & 0 & -2\mu & 0 & 0 \\ 0 & 0 & \mu & 0 & -\mu & 0 \\ 0 & 0 & 0 & 0 & 0 & 0 \end{bmatrix}. \tag{7.23}$$

The six states are defined as:

S_1 1 operating, 1 standby;
S_2 1 operating, 1 failed R;
S_3 1 operating, 1 failed NR;
S_4 2 failed R;
S_5 1 failed R, 1 failed NR;
S_6 2 failed NR.

Note that the last state, S_6, is absorbing. If in another phase of the mission, both items were required to be operating if available, but one would be enough, the matrix for that phase would be

$$Q_2 = \begin{bmatrix} -2\lambda_1-2\lambda_2 & 2\lambda_1 & 2\lambda_2 & 0 & 0 & 0 \\ \mu & -\mu-\lambda_1-\lambda_2 & 0 & \lambda_1 & \lambda_2 & 0 \\ 0 & 0 & -\lambda_1-\lambda_2 & 0 & \lambda_1 & \lambda_2 \\ 0 & 2\mu & 0 & -2\mu & 0 & 0 \\ 0 & 0 & \mu & 0 & -\mu & 0 \\ 0 & 0 & 0 & 0 & 0 & 0 \end{bmatrix}. \tag{7.24}$$

In both Q_1 and Q_2 S_4, S_5, and S_6 are downstates but only S_6 is irrecoverable. However, if two items were absolutely required then only S_1 would be an upstate and only S_2 and S_4 would be recoverable. Note that the states have the same meanings as to the states of the items of the system in both matrices.

The methods of Chapter 5 allow us to calculate mean time to or between defined events and their variances but do not directly help with finding the probability that a phased mission would be completed satisfactorily. Mission reliability as it is usually called, and mission availability (proportion of mission spent in satisfactory states) and other statistics are usually calculated by powering the probability matrix corresponding to the transition rate matrix. This is usually done by adding an identity matrix to Q, i.e. $(Q + I)$ and powering it up to the duration of the phase t, i.e. $(Q + I)^t$. This is then multiplied by the starting vector $\boldsymbol{P}_0$, usually $(1, 0, 0, 0, \ldots)$, to obtain an estimate of the end-of-first-phase state-probability vector $\boldsymbol{P}_1$, i.e.

$$\boldsymbol{P}_1 = \boldsymbol{P}_0(Q + I)^t. \tag{7.25}$$

$\boldsymbol{P}_1$ is then used as the starting vector for the next phase. If $\boldsymbol{P}_0 = (1, 0, 0, \ldots)$ then $\boldsymbol{P}_1$ is the top row of $(Q + I)^t$. However, this procedure actually involves the assumption that $(Q + I)$ approximates the matrix of conditional probabilities of transition in unit time.

This approximation is open to challenge once it has been clearly stated. Repeated powering of $(Q + I)$ until the matrix goes singular (all rows the same) does yield the same result as solving

$$Q^{\mathrm{Tr}}\boldsymbol{P} = (0), \quad \sum_{i=1}^{n} P_i = 1 \tag{7.26}$$

as set out in Chapter 5, but intermediate results for values of t less than, say, $3/\lambda$ will be in serious error for the following reasons.

1. The conditional probability of failure of an item in unit time is $1 - \exp(-\lambda)$ which may be close enough to λ but the corresponding repair rate is likely to be two orders greater and $\mu \approx 1 - \exp(-\mu)$ is usually a bad approximation.
2. In any finite unit of time such as an hour, more than one event can occur. It may be reasonable to neglect λ^2, but μ^2 and $\mu\lambda$ may be the same order or only an order smaller than μ and still an order or two bigger than λ. Reducing the unit of time also reduces the rates, but the relative position remains the same.

Also, the smaller the value of t, the greater is the relative error.

The alternative method of Laplace transforms, also given in Chapter 5 is not always available and is not amenable to computer solution. It is practical only for the very simplest systems. Returning to the matrix powering method, rewrite the matrix Q_2 as a matrix of conditional transfer probabilities in unit time, $Q_{2(\mathrm{p})}$, using

$$f_i = 1 - \mathrm{e}^{-\lambda_i}, \quad r = 1 - \mathrm{e}^{-\mu}, \tag{7.27}$$

where f_i and r are the probabilities of failure (repair) in unit time given the item is running (under repair) at the start of the unit of time

$$Q_{2(p)} = \begin{bmatrix} (1-f_1-f_2)^2 & 2f_1(1-f_1-f_2) & 2f_2(1-f_1-f_2) & f_1^2 & 2f_1f_2 & f_2^2 \\ r(1-f_1-f_2) & (1-r)(1-f_1-f_2)+rf_1 & rf_2 & f_1(1-r) & f_2(1-r) & 0 \\ 0 & 0 & 1-f_1-f_2 & 0 & f_1 & f_2 \\ r^2 & 2r(1-r) & 0 & (1-r)^2 & 0 & 0 \\ 0 & 0 & r & 0 & 1-r & 0 \\ 0 & 0 & 0 & 0 & 0 & 1 \end{bmatrix}. \tag{7.28}$$

System state definitions for (7.28) are identical to those for (7.24).

Now, although equation (7.28) looks as if it covers all probabilities, in fact this it not so. It omits the probability of double events (failure followed by repair or *vice versa*) in unit time to the same item. The probability of a failure followed by a repair in unit time is

$$p_{\text{fr}} = \int_0^1 \lambda e^{-\lambda_1 x} \mu e^{-\mu(1-x)} dx$$
$$= \lambda_1 \mu e^{-\mu}(e^{\mu-\lambda_1} - 1)/(\mu - \lambda_1). \tag{7.29}$$

Similarly, the probability of a repair followed by a failure in unit time is obtained by interchanging μ and λ_1, thus

$$p_{\text{rf}} = \mu\lambda_1 e^{-\lambda_1}(e^{\lambda_1-\mu} - 1)/(\lambda_1 - \mu) = p_{\text{fr}}. \tag{7.30}$$

Putting $\lambda_1 = 0.001$, $\mu = 0.1$ gives $p_{\text{fr}} = p_{\text{rf}} = 0.951 \times 10^{-4}$.

In practice, unless repairs are badly carried out and induce repeat failures, the real probability p_{rf} is much smaller than this calculation based upon Poisson processes would suggest. In the latter case the suggested solution is to count the event as one failure and one repair until it is fixed properly. This will tend to make both λ_1 and μ smaller, leaving availability unaffected. In the case of p_{fr}, recall that practical repair times usually fit a lognormal distribution better than an exponential, so that here again there is some justification for the assumption.

In constructing the probability matrices, even with the assumption that p_{fr} and p_{rf} are (close to) zero, it is still too easy to make mistakes. A systematic method for generating the matrices is therefore greatly to be desired.

In performing the powering, it is tempting to use established methods involving eigenvalues. However, because usually $r \gg f$ and f is small absolutely, the standard computer algorithms tend to be unstable. It is better to use double (or even triple) precision in a program in which Q_{p} is stored as a doubly subscripted variable and the powering is performed by nested DO loops. If this is done, the matrix can become singular within the accuracy limitations of the software without this leading to gross error as it does if diagonal forms and eigenvalues are used. It is also easy to incorporate in the program a switch to the steady-state solution should the matrix become singular, to avoid further useless calculation.

7.2.3 Method for phased missions

(a) Preconditions

A mission enters a new phase when the requirement for numbers and/or the method of operation (standby or active parallel) changes at a predetermined time since the last such change. Provided that the definitions of system states as so many up, so many down repairable, and so many down non-repairable, remain common through all phases of the mission, calculations can be made as outlined above and detailed below. It is *not* necessary that the number of items which are normally run should remain the same for all mission phases, *nor* that the number of simultaneous repairs possible remains the same from phase to phase.

(b) Method

We have seen already how the product of the starting vector and the probability matrix powered to the phase duration represents the vector of state probabilities at the end of the first phase. We can now generalize this to the calculation of the phase-end vector for the kth phase out of n phases

$$\boldsymbol{P}_{\mathrm{k}} = \boldsymbol{P}_{\mathrm{k}-1}\,(Q_{\mathrm{pk}})^{t_k}, \tag{7.31}$$

where t_k is the duration of phase k. We can also calculate the state probability vector $\boldsymbol{P}(T)$ at the end of any unit of time T from mission start within the kth phase, i.e. between T_{k-1} and T_k. By averaging these values over the interval we can estimate the expected time spent in each state i during phase k by averaging the $\boldsymbol{P}(T)$ over the interval

$$\bar{t}_{i,k} \approx [t_k/(t_k + 1)] \sum_{T=T_{k-1}}^{T_k} p_i(T). \tag{7.32}$$

Also the expected proportion of total time T_k spent in state i up to the end of phase k is

$$\bar{p}_i(T_k) = \sum_{j=1}^{k} \bar{t}_{i,j}/T_k. \tag{7.33}$$

By putting $k = n$, the overall mission figures can be obtained from (7.32) and (7.33).

(c) Number of failures in a mission

For logistical reasons it will be useful to know how many failures of each kind to expect during a mission. The expected or mean numbers of each type outstanding at the end of the mission can be found from $\boldsymbol{P}(T_k)$ as follows. Let $F_{1,i}$ and $F_{2,i}$ be the numbers of repairable and non-repairable failures which exist in, and incidentally define, state i, out of m states. Then the mean numbers of failures of each type outstanding at mission-end are

$$F_{1,k} = \sum_{i=1}^{m} p_i(T_k)\, F_{1,i} \quad \text{(repairable)}$$

$$F_{2,k} = \sum_{i=1}^{m} p_i(T_k)\, F_{2,i} \quad \text{(non-repairable).} \tag{7.34}$$

These two figures would give the repair facility an indication of the level of work to expect after the mission, particularly if more than one similar operational unit was involved. There will also be some repaired failures. The expected number of these per unit is

$$F_r = \mu \sum_{k=1}^{n} \sum_{i=1}^{m} \bar{t}_{i,k} g_{i,k}, \tag{7.35}$$

where $g_{i,k}$ is the number of repairs in progress in state i and phase k. In practice, $g_{i,k}$ is the lesser of the number of repair crews available and the number of extant repairable failures in state i, phase k. The total expected failures per mission is $F_r + F_{1,k}$ (repairable) plus $F_{2,k}$ (non-repairable).

(d) Standby systems

A convenient way to draw the transition rate diagrams for systems of items subject to two modes of failure is to track one mode horizontally and the other vertically. For example a two-out-of-three standby system with three repair teams would be drawn as in Fig. 7.10 below.

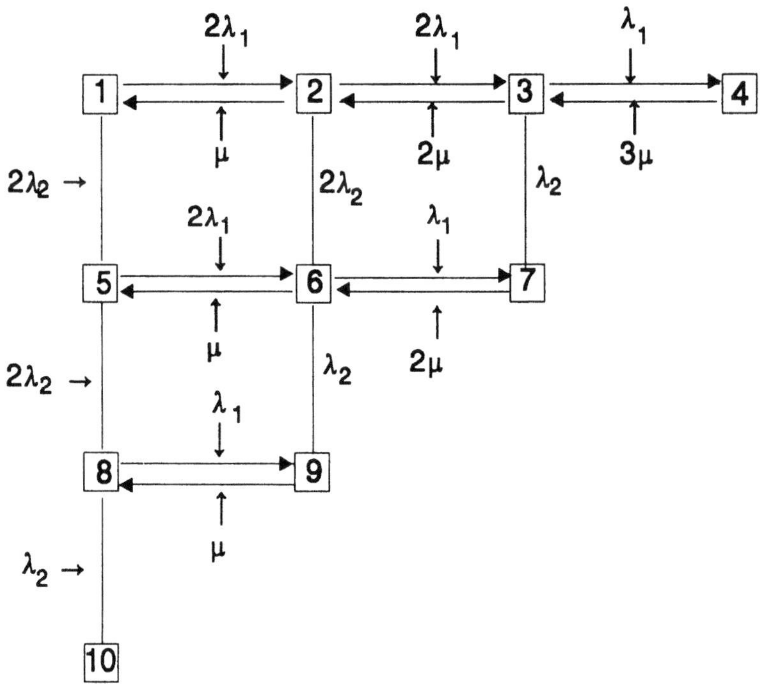

Fig. 7.10 State-space diagram for a two-out-of-three standby system with repairable and non-repairable failures, three repair teams and continue on system failure.

Figure 7.10 also provides a methodical way to generate the transition probability diagram. Thus from state 1 the transition rate diagram indicates two items at risk which may fail in either mode. Matrix entry $q_{1,1}$ is therefore $(1 - f_1 - f_2)^2$. To achieve S_2 one must fail repairable and the other remain intact with probability

$$2f_1(1 - f_1 - f_2).$$

Similarly,
$$q_{1,5} = 2f_2(1 - f_1 - f_2).$$

But two events may occur in the same hour so we have also to consider a direct entry to S_3 with probability f_1^2 or S_8 with probability f_2^2 or S_6 with probability $2f_1f_2$. To go from S_1 to S_7 in a single hour would involve the standby item failing in its first hour of operation. As this is likely to be its first hour of operation after repair in another phase or mission, this probability is ignored. This is a moot point; we could include it, but would have to overestimate its probability of failure or make special provision because it would not run for the full hour. It is simpler to ignore it, and this policy is also consistent with the other simplifying approximation. Similarly S_9 and S_{10} are not directly entered from S_2 but S_8 occurs with probability f_2^2. Taking S_2 next, all three items are involved, one is under repair and the other two are at risk of both modes of failure. The diagonal element is therefore $(1 - r)(1 - f_1 - f_2)^2 + 2rf_1$ which is the probability nothing fails or is repaired plus the probability of a failure to one and repair of another. All the things that can happen in unit time from S_1 can also occur from S_2 with the same probabilities multiplied by $(1 - r)$ to account for non-repair of the failure. In addition there are other probabilities multiplied by r. Take S_2 to S_7 for example. The transition rate diagram requires two steps via S_6 or S_3 which is two ways. So we have $2f_1f_2$, but as there is also a μ emanating from S_2 we must multiply by $(1 - r)$, giving $2f_1f_2(1 - r)$. As another example take S_6 to S_8. This involves a repair and an NR failure and there are two routes, via S_5 or S_9. As there is only one item operating in S_6 it cannot fail in both modes in the same hour so that λ_1 emanating from S_6 is ignored. The general rule is that there cannot be more steps in total than items operating and under repair, nor more failures than the number operating, nor more repairs than the number under repair. If the diagram shows x ways of proceeding from one state to another then there is a factor x in the probability, the others being f_i for λ and r for μ. As a check, the sum of each row of the matrix must be unity and the diagonal element is 1 minus the sum of the rest of the row. Also, the terms $(1 - f_1)^x$ and $(1 - f_2)^x$ can never occur because one mode of failure excludes the other.

(e) *Mission reliability*

The definition of mission failure, against which mission reliability is to be measured varies from phase to phase. In each phase, mission failure is

	1	2	3		5	6	7	8	9	10
1	$(1-f_1-f_2)^2$	$2f_1(1-f_1-f_2)$	f_1^2	0	$2f_2(1-f_1-f_2)$	$2f_1f_2$	0	f_2^2	0	0
2	$r(1-f_1-f_2)^2$	$(1-r)(1-f_1-f_2)^2+2rf_1$	$2(1-r)f_1(1-f_1-f_2)$	$(1-r)f_1^2$	$2rf_2(1-f_1-f_2)$	$2(1-r\lambda(1-f_1-f_2)f_2$	$2(1-r)f_1f_2$	f_2^2r	$(1-r)f_2^2$	0
3	$r^2(1-f_1-f_2)$	$2r(1-f_1-f_2)$	$(1-r)^2(1-f_1-f_2)+2rf_1$	$(1-r)^2f_1$	r^2f_2	$2rf_2$	$(1-r)^2f_2$	0	0	0
4	r^3	$3r^2(1-r)$	$3r(1-r)^2$	$(1-r)^3$	0	0	0	0	0	0
5	0	0	0	0	$(1-f_1-f_2)^2$	$2(f_1(1-f_1-f_2)$	f_1^2	$2f_2(1-f_1-f_2)$	$2f_1f_2$	f_2^2
6	0	0	0	0	$r(1-f_1-f_2)$	$(1-2)(1-f_1-f_2)+rf_1$	$2(1-r)f_1$	rf_2	0	0
7	0	0	0	0	r^2	$2r$	$(1-r)^2$	0	0	0
8	0	0	0	0	0	0	0	$1-f_1f_2$	f_1	f_2
9	0	0	0	0	0	0	0	r	$1-r$	0
10	0	0	0	0	0	0	0	0	0	1

Fig. 7.11 Transition probability matrix for a two-out-of-three standby system with identical items subject to repairable and non-repairable failures.

defined as being in certain states at the start of the phase or entering such a state during the phase. Other states may represent failure of the system but not of the mission. These will involve repairable failures. Such system failures are undesirable but not fatal to the mission. In other cases a state may be entered in an earlier phase, which is irrecoverable because of non-repairable failures, and which will be regarded as mission failure in a later phase. It is important that such conditions are regarded as mission failures straight away. Otherwise, logistical calculations will be in error to the extent that further failures, repairable or not, may occur. In practice should such failures occur, the mission would be abandoned at once and the unit would return to base. The mission of such a stricken unit would in fact change to one of getting safely back to base; this would be prejudiced by uselessly carrying on until unable to cope with a later phase of the original mission. So it is possible to have mission failure without system failure as well as *vice versa*.

It will be convenient to place the mission failure states at the right and bottom of the matrix. Let M_k be the first mission failed state in phase k such that states from $i = M_k$ to $i = m$ inclusive are mission-failed states. Now, the probability that the mission is still alive at the end of phase k is the probability of no fatal failure in phase k given no fatal failure in previous phases. We can find the probability of fatal failure in phase 1 by modifying the matrix to make all M mission-failed states absorbing, i.e. form matrix Q_{c1} by putting

$$\begin{aligned} q_{c1,i,j} &= q_{1,i,j} \quad \text{for } i,j < M_1 \\ q_{c1,i,j} &= 1 \quad \text{for } i \geqslant M_1 \\ q_{c1,i,j} &= 0 \quad \text{for } i \geqslant M_1,\ j \neq i. \end{aligned} \tag{7.36}$$

Then form the vector for the end of phase 1, $\boldsymbol{P}_{c1}$

$$\boldsymbol{P}_{c1} = \boldsymbol{P}_0(Q_{c1})^{t_1}. \tag{7.37}$$

The probability of failure in phase 1 is then

$$p_{f1} = \sum_{i=M_1}^{m} p_{c1,i}. \tag{7.38}$$

Then we can form the conditional starting vector for phase 2 of the mission given the mission is still alive at end of phase 1 by redistributing p_{f1} over the end-of-phase-1 upstates in proportion to their existing values, i.e.

$$\begin{aligned} p'_{c1,i} &= p_{c1,i}/(1-p_{f1}) \quad \text{for } 1 \leqslant i < M_1 \\ p'_{c1,i} &= 0 \qquad\qquad\qquad\;\; \text{for } i \geqslant M_1. \end{aligned} \tag{7.39}$$

The procedure is then repeated for the second and subsequent phases. The mission reliability is given by

$$R_m = 1 - \sum_{i=1}^{m} p_{fi}$$

(f) Mission availability A_m

This may be defined as the expected proportion of mission time spent in states regarded as satisfactory for the phase. The implication is that there are recoverable states which in specified phases are regarded as system failed (temporarily) but not mission failed. In other phases, mission and system failure are likely to be synonymous. For example, in the action phase of a weapon delivery system's mission there is no time to fix any outstanding repairable failures and so it is the total number of failures extant rather than their nature which determines mission reliability. They must be up at the time of actual delivery even if they (or some) were allowed to be temporarily failed during previous phases

$$A_m = \sum_{k=1}^{n} \sum_{i=1}^{M_k - 1} \bar{t}_{i,k}/T_m \tag{7.40}$$

where $\bar{t}_{i,k}$ is defined by equation (7.32). Equation (7.44) gives the proportion of system uptime in successful missions i.e. given that the mission goes through to the end of phase m. Another definition of A_m would give the ratio of expected achieved system uptime to the planned system use time for the mission T_m, thus

$$A'_m = \sum_{k=1}^{n} \sum_{i=1}^{M_k - 1} \left[\bar{t}_{i,k}\left(1 - \sum_{j=1}^{k-1} p_{fj}\right)\right]/T_m, \tag{7.41}$$

where p_{fj} is defined as in equation (7.38). A'_m has $\bar{t}_{j,k}$ conditional upon entering phase k; thus the terms for the kth and later phases become zero if that phase is not entered. A'_m combines A_m and R_m into a single measure.

Notice that both R_m and A_m refer to the ultimate outcome of the mission and the equations defining these do not help much with calculations of $R(T)$ or $A(T)$ or $\bar{A}(T)$ the reliability and instantaneous and average availabilities to a time T into the mission.

7.2.4 Formation of matrices

(a) Number of states

The first requirement is to know how many states there can be for a system of n identical items each of which may be in any of s states. It turns out that the number is a simple combinatorial, thus

$$N_{s,n} = \binom{n+s-1}{s-1} = \binom{n+s-1}{n}. \tag{7.42}$$

Thus, for example, if $n = 4$, $s = 3$, then $N_{3,4} = \binom{6}{2} = 15$ states or, in general, the numerical coefficient of terms of the binomial expansion of power n, which can be obtained from Pascal's triangle.

Notice that if the items are non-identical, the number of states is s^n, which quickly becomes very large indeed as s and n increase. Because every computer has an upper limit of size of matrix or doubly subscripted variable which can be handled, it has become customary to use simulation methods for complex systems of unlike items even when $s = 2$ only. However we have seen in earlier chapters that this is not always necessary. To a certain extent, this resort to simulation was caused by failure to consider or develop analytical methods of calculation other than those involving matrices.

(b) Kronecker algebra

For the next step in the formation of matrices we need to define what is meant by a Kronecker product of two matrices, thus

$$A \otimes B = (a_{ij} \times B). \tag{7.43}$$

This is to say that if A is a $m \times n$ matrix and B is a $p \times q$ matrix, $A \otimes B$ will have dimensions $mp \times nq$. The Kronecker sum of two matrices is given by

$$A \oplus B = A \otimes I_b + I_a \otimes B. \tag{7.44}$$

Note the following properties for Kronecker matrix algebra

1. $A \otimes B \neq B \otimes A$
2. $A \oplus B \neq B \oplus A$

3. $(A \otimes B) \otimes C = A \otimes (B \otimes C)$
4. $(A \oplus B) \oplus C \neq A \oplus (B \oplus C)$.

It can also be shown that the transition-rate matrices for a system can be built up by forming successive Kronecker sums, and probability matrices by forming the Kronecker products of their individual matrices. For example, the probability matrix for a single item with R and NR failures is

$$Q_{\mathrm{p}} = \begin{bmatrix} 1 - f_1 - f_2 & f_1 & f_2 \\ r & 1 - r & 0 \\ 0 & 0 & 1 \end{bmatrix}. \tag{7.45}$$

The Kronecker square of this matrix, written $Q^{\otimes 2}$, is

$$\begin{bmatrix} (1-f_1-f_2)^2 & f_1(1-f_1-f_2) & f_2(1-f_1-f_2) & f_1(1-f_1-f_2) & f_1^2 & f_1f_2 & f_2(1-f_1-f_2) & f_1f_2 & f_2^2 \\ r(1-f_1-f_2) & (1-r)(1-f_1-f_2) & 0 & f_1^2 & f_1(1-r) & 0 & rf_2 & (1-r)f_2 & 0 \\ 0 & 0 & 1-f_1-f_2 & 0 & 0 & f_1 & 0 & 0 & f_2 \\ r(1-f_1-f_2) & rf_1 & rf_2 & (1-r)(1-f_1-f_2) & f_1(1-r) & f_2(1-r) & 0 & 0 & 0 \\ r^2 & r(1-r) & 0 & r(1-r) & (1-r)^2 & 0 & 0 & 0 & 0 \\ 0 & 0 & r & 0 & 0 & 1-r & 0 & 0 & 0 \\ 0 & 0 & 0 & 0 & 0 & 0 & 1-f_1-f_2 & f_1 & f_2 \\ 0 & 0 & 0 & 0 & 0 & 0 & r & 1-r & 0 \\ 0 & 0 & 0 & 0 & 0 & 0 & 0 & 0 & 1 \end{bmatrix}. \tag{7.46}$$

By rearranging and combining rows and columns within the rules of (ordinary) matrix algebra this matrix becomes the same as equation (7.28). Combine rows and columns 2 and 4, 3 and 7, and 6 and 8, to get the up and repairable states to the top and left. The secret is to combine rows and columns which have identical diagonal elements.

The combination of transition-rate matrices by taking the Kronecker sum is left as an exercise for the reader. Despite property number 2 above order is not important ultimately; if the matrices are square and of the same size they can be rearranged to correspond with one another. Notice that in either case the resulting matrix has dimensions s^n-before reduction, since the method is 'blind' to the fact that the items are identical. The Kronecker product or sum methodology will of course give the correct matrix if applied to non-identical items. However it is restricted to cases in which all items operate or are under repair until they fail NR. This does not prevent higher states than 'all items down NR' being regarded as 'system failed'' or 'mission failed' but the assumption in this method of forming the matrix is that the system carries on to the bitter end with whatever is still serviceable or repairable.

Matrix powering is substantially simplified if n machines are operated as above. This is because it can be shown that the Kronecker product of the powered single-machine matrices is equivalent to the powered Kronecker product of the basic matrices, i.e.

$$(Q^{\otimes n})^t \equiv \Pi^{\otimes n}(Q^t) \text{ for identical machines,} \tag{7.47}$$

or, more generally,

$$\left(\prod_{i-1}^{n\,\otimes} Q_i\right)^t \equiv \prod_{i-1}^{n\,\otimes} (Q_i^t) \text{ for non-identical machines,} \qquad (7.47a)$$

where $\Pi^{\otimes}$ denotes a serial Kronecker product and $\Pi^{\otimes n}$ the nth Kronecker power of a matrix.

7.2.5 General analysis of three-state items

(a) Introduction

In his critique of the paper by Jenney and Sherwin (1986), Malon pointed out that if more than one path through a system of items subject to open and short failures is required or more than two or more items in physical series are needed for satisfactory operation then terms in $(1 - q - s)$ and its powers would appear in the solutions. It actually does not matter how the three item states are differentiated, provided that they are exhaustive. The mathematics remains the same. The open- and short-circuit formulae in the early part of this chapter were found by simple probability, but these involved only a minimum of one item required to remain working and connected to both end-nodes of the system.

EXAMPLE 7.4

Consider the case of a system of two identical items each of which can fail in two modes into probabilities q and s (which may or may not correspond to open and short). The probabilities q and s may be the probabilities over a fixed time of interest or one-shot probabilities. We can proceed in exactly the same way as for the construction of the top line of the matrix (7.28) above. The probabilities of the various possible states of the system after operation are given in Table 7.1.

If the system is actually a physically parallel arrangement then if only one path is required S_1 and S_2 are upstates and S_3 to S_6 inclusive are downstates. Adding together either p_1 and p_2 or p_3 to p_6 inclusive gives the same result in which the terms in $(1 - q - s)$ and qs cancel out, giving

Table 7.1 Possible outcomes for a two item system of three-state items

State	*Description*	*Probability p_i*
S_1	Both items succeed	$(1 - q - s)^2$
S_2	One fails open	$2q(1 - q - s)$
S_3	Both fail open	q^2
S_4	One fails short	$2s(1 - q - s)$
S_5	Both fail short	s^2
S_6	One fails open, the other short	$2qs$

$R = (1 - s)^2 - q^2$, as before. If the items operate in series then S_1 and S_4 are the upstates and $R = (1 - q)^2 - s^2$, as before. However, if both items must operate for system success, then (in this case regardless of physical arrangement) the reliability is $(1 - q - s)^2$.

Generalizing to j paths (connected series items) required and m items in parallel (series), the unreliability expression will consist of the first $m - j + 1$ terms involving open (short) failures plus all the terms involving short (open) failures. For parallel systems, j paths we have therefore

$$R_{mpj} = (1 - s)^m - \sum_{i=0}^{j-1} \binom{m}{i} q^{m-i}(1 - q - s)^i \tag{7.48}$$

and for series systems of n items with k items required

$$R_{nsk} = (1 - q)^n - \sum_{i=0}^{k-1} \binom{n}{i} s^{n-i}(1 - \mathrm{q} - \mathrm{s})^i. \tag{7.49}$$

Extension of the theory to cover series parallel and parallel series arrays in which j paths (each with at least k working items) is left as an exercise for the reader.

REFERENCES

BS 5760 (1981), Reliability of Systems Equipments and Components, Part 2, Guide to the Assessment of Reliability, British Standards Institution, London.

Brucker, J. (1986) (unpublished) obtainable from author at University of California, Los Angeles, California 90024 USA.

Dhillon, B.S. and Singh, C. (1981) *Engineering Reliability – New Techniques and Applications*, Wiley, pp. 329–34.

Jenney, B.W. and Sherwin, D.J. (1986) *IEEE Trans. Rel.*, **R35**(5), 532–8.

Lloyd, D.K. and Lipow, M. (1964) *Reliability Management Methods and Mathematics*, Prentice Hall, NJ, pp. 243–8.

Malon, D.M., Comments on Jenney and Sherwin (1986).

Shooman, M.L. (1968) *Probalistic Reliability: An engineering approach*, McGraw Hill, NJ.

Von Alven, W. (ed.) (1965) *Reliability Engineering*, Prentice Hall, NJ.

CHAPTER 8

Boolean algebra and probability laws for reliability evaluations

Often the reliability or availability of large systems can only be described by complex logical functions of basic events such as component up and down states. In the special case that each basic event appears only once in the logical expression and all basic events in the expression are independent, then the probability evaluation is simple and straightforward since the probability can be calculated term by term. For example for the probability of the logical expression in equation (8.1) the probabilities x, y and z of the terms between brackets are evaluated first after which the overall probability is evaluated from the reduced expression in X, Y and Z (equation (8.2)).

$$\left.\begin{aligned} S &= (A \cap B) \cup (C \cup D) \cap (\bar{E} \cup F) \\ &= X \cup (Y \cap Z), \end{aligned}\right\} \quad (8.1)$$

where $\cap$ implies intersection, (i.e. both A and B), $\cup$ stands for union (i.e. either A or B or both), and $\bar{A}$ means not A.

$$Pr(S) = x + (yz) - x(yz)$$

with

$$\left.\begin{aligned} x &= ab \\ y &= c + d - cd \\ z &= \bar{e} + f - \bar{e}f. \end{aligned}\right\} \quad (8.2)$$

For other cases the logical expressions usually have to be rewritten to a simpler form that facilitates probability calculations. This can be done by using Boolean algebra.

8.1 BOOLEAN ALGEBRA

The Boolean algebra allows one to change logical expressions to more convenient forms for probability calculations. A total of three different basic logical operations can be distinguished: inversion, intersection (logic product) and union (logic sum) which have to be applied in that order unless the order is changed by brackets. This is just like in normal algebra. Furthermore the following laws are applicable to logic functions.

The equality laws

$$\left.\begin{aligned} A \cup A &= A \\ A \cap A &= A \end{aligned}\right\} \tag{8.3}$$

This law says that the intersection or the union of identical events is the event itself and it prevents multiple counting of events in calculations.

The associative laws

$$\left.\begin{aligned} A \cup (B \cup C) &= (A \cup B) \cup C \\ A \cap (B \cap C) &= (A \cap B) \cap C \end{aligned}\right\} \tag{8.4}$$

which state that the order in which identical operations are performed is not important. The Venn diagrams for these laws can be found in Fig. 8.1.

The commutative laws

$$\left.\begin{aligned} A \cup B &= B \cup A \\ A \cap B &= B \cap A \end{aligned}\right\} \tag{8.5}$$

which states that the order of appearance of the terms in a logical union or intersection is irrelevant as can be seen from Fig. 8.2.

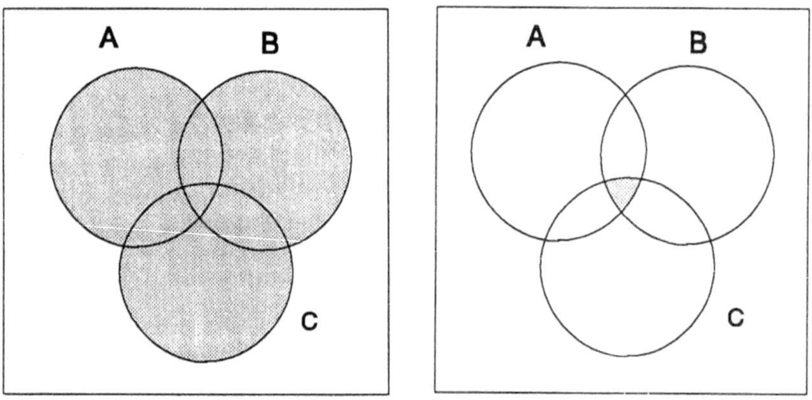

Fig. 8.1 Venn diagrams showing associative laws for logical union and logical intersection.

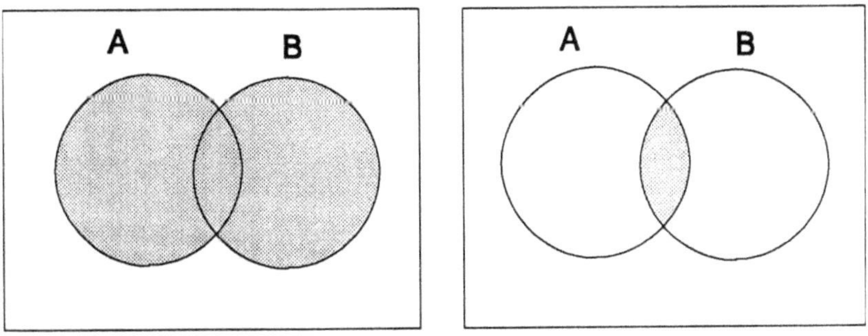

Fig. 8.2 Venn diagrams to highlight the commutative laws for logic union and logic intersection.

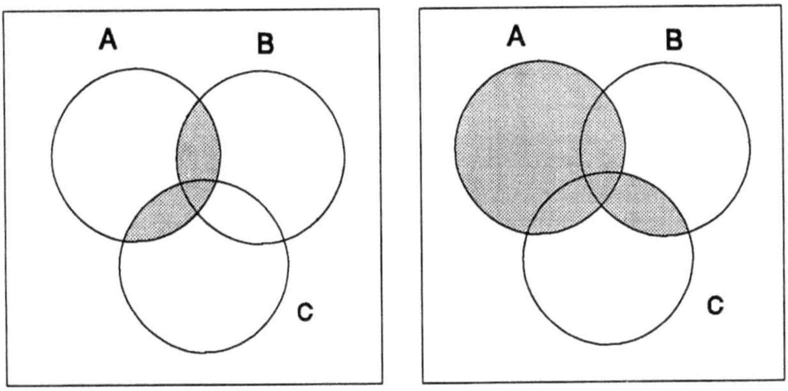

Fig. 8.3 Venn diagrams to illustrate the distributive laws in equation (8.6).

The distributive laws

$$\left.\begin{aligned} A \cap (B \cup C) &= (A \cap B) \cup (A \cap C) \\ A \cup (B \cap C) &= (A \cup B) \cap (A \cup C) \end{aligned}\right\} \tag{8.6}$$

The Venn diagrams for these equations are depicted in Fig. 8.3.

The negation laws

$$\left.\begin{aligned} A \cup \bar{A} &= 1 \text{ (universal event)} \\ A \cap \bar{A} &= 0 \text{ (empty event)} \\ \bar{\bar{A}} &= A \end{aligned}\right\} \tag{8.7}$$

The de Morgan's laws

$$\left.\begin{aligned} \overline{A \cup B} &= \bar{A} \cap \bar{B} \\ \overline{A \cap B} &= \bar{A} \cup \bar{B} \end{aligned}\right\} \tag{8.8}$$

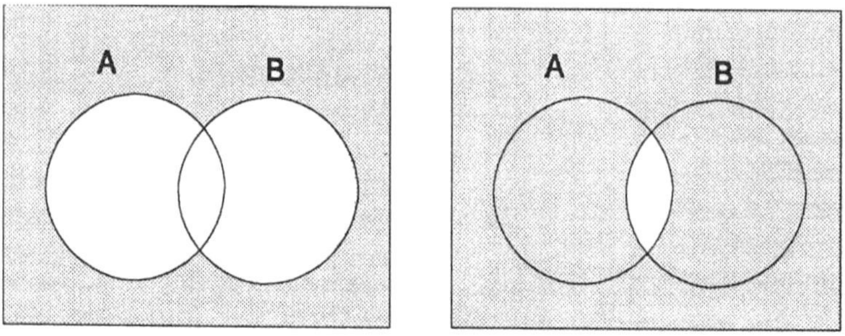

Fig. 8.4 Venn diagrams to illustrate the de Morgan rules.

These laws imply that the inverse of a union (intersection) of events is identical to the intersection (union) of the inverse events (Fig. 8.4). These laws are very helpful in finding the dual (inverse) form of a logical expression.

The laws above also hold for systems with multi-state items, where it should be noted that for an n-state item

$$\left.\begin{aligned} \bar{A}_i &= \bigcup_{\substack{j=0 \\ j\neq i}}^{n} A_j \\ \text{and} \\ A_i \cap A_j &= 0 \end{aligned}\right\} \tag{8.9}$$

because all states of an item are mutually exclusive by definition.

Most textbooks on reliability adopt a simplified notation for Boolean expressions in which the set symbols for union and intersection $\cup$ and $\cap$ are replaced by $+$ and $\cdot$, respectively. We will adopt the same notation in the remainder of this book. However, in order to prevent any misunderstanding, the events in Boolean expressions will be written in capitals while in algebraic expressions their probabilities will be represented by small characters or as a probability function *Pr*( ).

8.2 PROBABILITY EVALUATION OF LOGICAL EXPRESSIONS WITH DEPENDENT AND INDEPENDENT EVENTS

The Boolean algebra introduced above is valid for both *dependent* and *independent* events. Probability calculations, however, are seriously complicated by dependencies of which the impacts on the probability are not well understood. Although Bayes' theorem allows us to express the probabilities of logic union and logic intersection as:

$$\left.\begin{aligned} Pr(A \cap B) &= Pr(A) \cdot Pr(B|A) = Pr(B) \cdot Pr(A|B) \\ Pr(A \cup B) &= Pr(A) + Pr(B) - Pr(A \cap B) \end{aligned}\right\} \quad (8.10)$$

the enumeration is strongly hampered when the conditional probabilities $Pr(B|A)$ and $Pr(A|B)$, are not known. In practice two types of dependencies can be distinguished especially:

1. common cause failures;
2. secondary failures.

Common cause failures occur when a single event affects the successful functioning of multiple parts in the system. Examples of such events are: the failure of a common power supply causing multiple machines to stop; or severe weather conditions increasing the failure probabilities of all overhead electricity distribution cables in a wide area.

Secondary failures occur when a failure of an item (primary failure) causes excess stress on one or more neighbour items causing them to break down more rapidly (secondary failures). An example of these items are electrical generators in active parallel redundancy. Normally the load is shared equally between the generators, however, when one fails two things may happen: the increased load on the second generator causes increased wear-out; a short circuit failure in the first generator causes a severe overload situation leading to an (almost) immediate breakdown of the second generator. It will be clear that in the latter case a serious design error has been made.

When the impacts of the originating events are direct and clear, the conditional probabilities usually are easy to assess, e.g. in the case of the common power supply failure above, the corresponding conditional failure probabilities of the connected machines are 1 (complete dependency). These dependencies can easily be coped with in probability calculations when all failures are analysed down to the cause. In that case the common cause appears multiple times in the Boolean expression as an independent event.

In other cases it is more difficult to quantify the (partial) dependency since quantification requires a large amount of failure data. This is almost impossible since even for the quantification of the failure rate of a single item it is very difficult to gather sufficient failure data let alone for two items in combination. Therefore, in practice one is satisfied usually with the knowledge that the possibility of interference has been brought down as much as possible.

Even when it is possible to quantify the dependency as an increased failure rate, the Boolean expression is not the most suitable tool for reliability evaluations on the system since the time delays between primary events and secondary events cannot be modeled with simple logical

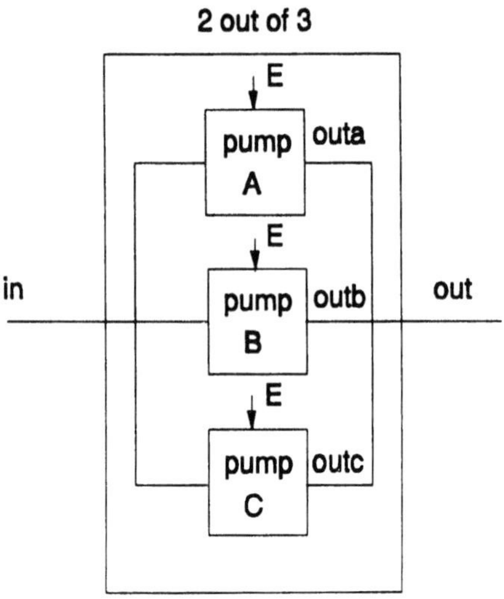

Fig. 8.5 Diagram of the two-out-of-three redundant pump unit.

expressions. In these cases Markov methods (Chapter 5) are more appropriate.

EXAMPLE 8.1

A small town is supplied with drinking water by means of a pipeline. The flow through this pipe is sustained by three pumps in active two-out-of-three redundancy (Fig. 8.5). All pumps are driven electrically from the same power supply. The Boolean expression for insufficient flow can be expressed in terms of the pumps' output flows which again can be split into independent basic events

$$\left.\begin{aligned} F &= \overline{outa} \cap \overline{outb} \cup \overline{outa} \cap \overline{outc} \cup \overline{outb} \cap \overline{outc} \\ &= (\bar{A} \cup \bar{E}) \cap (\bar{B} \cup \bar{E}) \cup (\bar{A} \cup \bar{E}) \cap (\bar{C} \cup \bar{E}) \cup (\bar{B} \cup \bar{E}) \cap (\bar{C} \cup \bar{E}), \end{aligned}\right\} \quad (8.11)$$

where A, B and C are the respective pumps and E is the power supply and $\overline{outi}$ means the absence of outputflow from pump I. For probability evaluation, Boolean expressions are usually rewritten in the more convenient sum-of-products form. With the help of the distributive, associative and equality laws from the Boolean algebra the sum-of-product form for expression (8.11) can be found as

$$F = \bar{E} + \bar{A} \cdot \bar{B} + \bar{A} \cdot \bar{C} + \bar{B} \cdot \bar{C}.$$

8.2.1 Inclusion-exclusion rule

Probability evaluation on a term-by-term base as was shown in formula (8.1) is not possible since the multiple appearance of identical events causes the different terms to be dependent. A more general solution that is able to deal with repeated events and also with mutually exclusive events (e.g. two different states of an item) is the so-called inclusion-exclusion rule which proceeds as follows:

1. *Calculate and add the probabilities of all single terms (products).* This however means that the probability of the intersection of any two terms X_i and X_j is added twice – once in the probability of X_i and once as part of the probability of X_j – and should be subtracted once again.

2. *Subtract the probabilities of all intersections of two terms from the previous result.* The exact result still has not been reached since the contribution of the intersection of any three terms X_i, X_j and X_k has been added three times through the single terms and subsequently has been subtracted three times again through the three twofold intersections of these terms. Hence, the probabilities of all threefold intersections of terms still have to be added.

3. *Repeat step two for all higher order intersections of terms, where it must be noted that odd combinations of terms have positive contributions to the final result while the combinations of an even number of terms have negative contributions.*

In formula the inclusion-exclusion rule for a sum-of-product expression with n terms looks like

$$\left.\begin{aligned} Pr\{F\} = &+\sum_{i=1}^{n} Pr\{X_i\} && \text{single terms} \\ &-\sum_{\substack{i=2\\ i\neq j}}^{n-1} Pr\{X_i \cap X_j\} && \text{double intersections} \\ &\vdots \\ &+(-1)^{n+1} Pr\{X_1 \cap X_2 \cap \ldots \cap X_n\} && n\text{th-order intersection} \end{aligned}\right\} \quad (8.12)$$

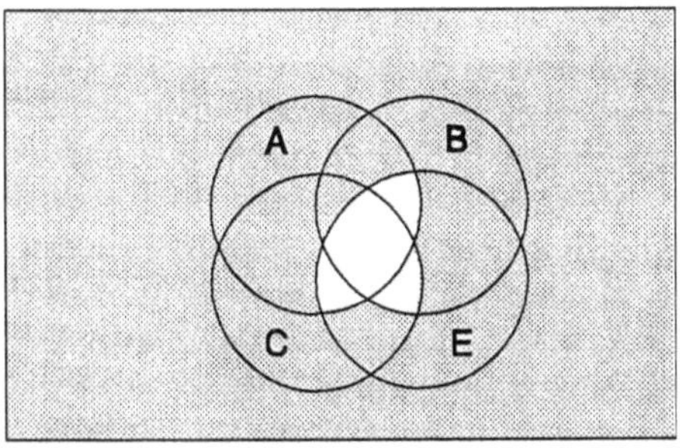

Fig. 8.6 Venn diagram for the water supply system of Example 8.2. The shaded area shows the failure mode of the system.

EXAMPLE 8.2

For the redundant water supply system of Example 8.1 the failure probability can now be calculated as

$$\left.\begin{aligned} Pr\{F\} = &+ \{\bar{e} + \overline{ab} + \overline{ac} + \overline{bc}\} \\ &- \{\overline{abe} + \overline{ace} + \overline{bce} + 3\cdot\overline{abc}\} \\ &+ \{3\cdot\overline{abce} + \overline{abc}\} \\ &- \overline{abce}. \end{aligned}\right\} \tag{8.13}$$

The Venn diagram in Fig. 8.6 illustrates this failure mode.

The inclusion-exclusion method always yields the exact answer after the calculation of the nth order intersections. However, it is a very tedious and time-consuming calculation especially when a large number of terms has to be considered. In some cases it might even be impossible to evaluate the exact solution due to limitations on computing power. In fault-tree analysis (Chapter 9) with all low-probability (rare) events the inclusion-exclusion rule converges very rapidly to the exact answer and a first- or second-order approximation usually suffices. However, in communication networks, where a low load of the edges (links) is economically unattractive, neither the edge availabilities nor the edge unavailabilities are so small. In this case the inclusion-exclusion calculation might even diverge initially before converging to the exact answer after the n-th order terms.

8.2.2 Disjoint-term method

Another method that will lead to an exact answer much faster is by making all terms in the sum-of-product expression mutually exclusive or disjoint so that the term probabilities can simply be added to find the exact answer. The procedure to do this is simple and straightforward:

1. Rank the terms according to increasing size.
2. Make the second term disjoint to the first term. This will yield one new term for each event in term 1 that doesn't appear in term 2, e.g.

$$\begin{aligned} &\text{term 1} \quad AXYZ \\ &\text{term 2} \quad APQR \rightarrow APQR\bar{X} \\ &\qquad\qquad\qquad\quad \rightarrow APQRX\bar{Y} \\ &\qquad\qquad\qquad\quad \rightarrow APQRXY\bar{Z} \end{aligned}$$

3. Make each other term disjoint to all its preceding terms in the same manner.

For the redundant water pump station above, each term lacks only one event with respect to each preceding term. In this special case the number of disjoint terms equals the number of original terms. The disjoint set of terms for this example is:

term 1 $\bar{E}$
term 2 $\bar{A}\bar{B} \rightarrow \bar{A}\bar{B}E$
term 3 $\bar{A}\bar{C} \rightarrow \bar{A}\bar{C}EB$
term 4 $\bar{B}\bar{C} \rightarrow \bar{B}\bar{C}EA$.

The probability of pump station failure now can be calculated as

$$Pr\{F\} = \bar{e} + \bar{a}\bar{b}e + \bar{a}b\bar{c}e + a\bar{b}\bar{c}e. \tag{8.14}$$

Comparing equations (8.13) and (8.14) it can be seen that the number of sub-calculations necessary to find the exact answer has been reduced from 15 to only 4. Therefore, it can be concluded that the disjoint-term method has large advantage over the inclusion-exclusion method when the event probabilities are not small enough to ensure rapid convergence of the inclusion-exclusion method.

8.3 CUT AND TIE SETS, MINIMAL CUTS AND MINIMAL TIES

The terms in a sum-of-products expression are often referred to as cuts or ties depending on the nature of the overall event described. When the expression describes system failure the corresponding terms are referred to as cuts while the terms are called ties when successful system performance is described. The set of all terms in an expression is then referred to as a cut set or tie set. This terminology originally stems from the network reliability field where a cut is a combination of failed edges that ensures the disconnection (cut) of the network while a tie is a combination of successful edges that ensures the connection between two vertices of the network. Nowadays both terms are used more generally as described above. Sometimes, the terms are also more generally referred to as *implicants*, especially when the expression neither describes clear system failure nor clear system success.

When all events in a cut are relevant, i.e. no event can be removed without losing the guarantee of system failure the cut is a minimal cut. Likewise, a tie in which all events are relevant is a minimal tie. A complete set of minimal terms (cuts or ties) is a sufficient basis for probability calculations. Any additional non-minimal term does not add to the probability and can be left out.

8.4 CONCLUSIONS

It has been shown that the Boolean algebra can be used to rewrite complex logical expressions that often appear in reliability analysis, into a sum-of-product form which is more suitable for probability evaluations.

The product terms in these expressions are called cuts or ties depending on whether they describe system failure or system success. The Boolean algebra applies to both dependent events and independent events, while multistate items can be considered as well.

Although Bayes' theorem allows one to express the probability of a logic function including dependent events, the enumeration of this probability is possible only when the conditional probabilities of the dependent events are known. However, when a logic expression can be rewritten to the sum-of-product form with no other dependencies than mutually exclusive events and repeated events, the probability can be evaluated by two methods:

1. the inclusion-exclusion method which quickly converges to the exact answer when the probabilities of the basic events are small;
2. the disjoint-term method which is more suitable when the basic probabilities are too large to ensure quick convergence of the inclusion-exclusion method.

8.5 WORKED PROBLEMS

PROBLEM 8.1

Use the Boolean algebra rules to rewrite the following expressions to their simplest sum-of-product form

$$S_1 = (A + B) \cdot (A + \bar{B} + \bar{C}) \cdot (\bar{A} + B + C) \cdot (\bar{B} + C)$$

$$S_2 = (A \cdot (B + C) + A\bar{B}\bar{C}) \cdot (A + C).$$

ANSWER

$$\begin{aligned} S_1 &= (A + B) \cdot (A + \bar{B} + \bar{C}) \cdot (\bar{A} + B + C) \cdot (\bar{B} + C) \\ &= (A + \bar{B}\bar{C}) \cdot (\bar{A}\bar{B} + C) \\ &= AC. \end{aligned}$$

$$\begin{aligned} S_2 &= (A \cdot (B + C) + A\bar{B}\bar{C}) \cdot (A + C) \\ &= A(B + C + \bar{B}\bar{C}) \\ &= A((B + C) + (\overline{B + C})) \\ &= A. \end{aligned}$$

PROBLEM 8.2

Apply the de Morgan rules to find the inverse function of $S = AB + \bar{A}C + A\bar{B}\bar{C}$ and rewrite it into a sum-of-product form.

ANSWER

$$\begin{aligned} \bar{S} &= \overline{AB + \bar{A}C + A\bar{B}\bar{C}} \\ &= \overline{AB} \cdot \overline{\bar{A}C} \cdot \overline{A\bar{B}\bar{C}} \end{aligned}$$

$$= (\bar{A} + \bar{B}) \cdot (A + \bar{C}) \cdot (\bar{A} + B + C)$$
$$= \bar{A}\bar{C} + A\bar{B}C.$$

PROBLEM 8.3

Evaluate the probability of the logical function $S = AB + AD + A\bar{B}\bar{C}$ with help of the inclusion-exclusion rule.

ANSWER

$$Pr(S) = ab + ad + a\bar{b}\bar{c} - abd - a\bar{b}\bar{c}d$$

PROBLEM 8.4

A smoke detections circuit of a fire-alarm system may fail in two manners. On the one hand the circuit may fail to detect (and signal) the presence of smoke (with probability 0.05) and on the other hand a false signal may be given in the absence of smoke (also with probability 0.05).

Obviously, this circuit is not very reliable and, therefore, it is decided to apply three of these circuits in active two-out-of-three redundancy, i.e. a warning signal is given when two or more circuits indicate smoke.

1. Derive the logical expression for system failure in terms of the single units.
2. Calculate the probability of failure.

ANSWER

1. The system is unable to detect smoke either when two units give a false signal or when two units fail to detect smoke. The other possibilities of two units being failed, i.e. one generating a false 'smoke detect' signal and one unable to detect smoke, do not constitute system failure since the signal of the good unit is still decisive. The logical expression for system failure is then

$$F = A_1B_1 + A_1C_1 + B_1C_1 + A_2B_2 + A_2C_2 + B_2C_2.$$

2. The inclusion-exclusion-exclusion rule would require $2^n - 1 = 63$ terms to be calculated to achieve an exact answer. In this case where the number of independent items is low (3) the disjoint-term method is favourable. The disjoint form of F requires only 10 calculations

$$F = A_1B_1 + A_1(B_0 + B_2)C_1 + (A_0 + A_2)B_1C_1 + A_2B_2$$
$$+ A_2(B_0 + B_1)C_2 + (A_0 + A_1)B_2C_2$$

Then

$$Pr(F) = 2 \cdot 0.05^2 + 4 \cdot 0.95 \cdot 0.05^2 = 0.0145.$$

However, a first-order approximation with the inclusion-exclusion rule requires only six terms and yields

$$Pr(F) \approx 6 \cdot 0.05^2 = 0.015$$

which, in this case, is a reasonable approximation.

REFERENCES

Dhillon, B.S. and Singh, C. (1981) *Engineering Reliability – New Techniques and Applications*, Wiley, New York, pp. 329–34.

Feller, W. (1968) *An Introduction to Probability Theory and its Applications*, Vol. I, 3rd edn, Wiley, New York, pp. 98–101.

Hohn, F.E. (1966) *Applied Boolean Algebra*, Macmillan, New York.

Mendelson, E. (1970) *Boolean Algebra and Switching Circuits*, McGraw-Hill, New York.

Shooman, M.L. (1968) *Probabilistic Reliability: An Engineering Approach*, McGraw-Hill, New York.

CHAPTER 9

Fault trees, event trees and success trees

Fault-tree analysis was developed by H.A. Watson of the Bell Telephone Laboratories in 1961–2 as part of a US Air Force study contract for the Minuteman missile launch control system. Through the years it has proved to be a very valuable tool for the reliability evaluation of complex systems, such as nuclear power stations, chemical plants, wide-body aeroplanes and communication satellites. Fault-tree analysis is a deductive process that splits up a complex event (system failure) into more detailed events, such as subsystem failures, that might be responsible for this occurrence. This process is repeated for each new event found and proceeds until only basic events lacking more detailed descriptions are left. Since fault tree analysis apparently uses a top-down approach, descending from the system level to more detailed levels like subsystem or component levels, as is also the case in a system design process, it is well suited to evaluate the reliability effects of design considerations at each stage of the design process. This, and the fact that one is free to choose the level of detail to be reached, has made fault tree analysis very popular. Analogous to fault trees, which describe undesired system performance (failure), success trees can be constructed describing the successful action of a system. The theory in this chapter applies to both types of event trees. Although the general term 'event tree' is more correct we will stick to the current term 'fault tree' since reliability analysts are particularly interested in system failures because these are the events of which the occurrences and impacts must be minimized.

Over the last two decades many papers and books dealing with fault-tree analysis have been published. Most of this literature is confined to systems having only two-state components. This means that all components are considered to be either good or failed and no distinction is made between different failure modes. Practical systems, however, may contain components having multiple failure mechanisms, and each mechanism may have a different influence on the system performance, thus leading to distinct failure modes. Furthermore, a sufficient description of system performance might require multiple degrees of component deterioration caused by a single failure mechanism to be distinguished. These multiple failure modes might introduce dependent events in a fault tree which will complicate the solution. Therefore, this chapter will show both; a simplified solution which can be applied to fault trees with all independent events and a formal solution method which has general validity.

Likewise, a system might have multiple failure modes as well. Therefore, a fault tree can be set up for each of a system's failure modes or each combination of these failure modes. The failure mode (or union of failure modes) considered is called the top event of the corresponding fault tree. Few authors considered these multi-state components (and systems) when introducing algorithms to calculate the probability of a fault tree's top event.

9.1 FAULT-TREE SYMBOLS AND DEFINITIONS

A fault tree is an oriented graph representing the causal relationships between the top event, i.e. the complex event to be analysed, and its causes. The name of the graph is derived from its form, which is very similar to that of a tree (Fig. 9.1). The symbols used in fault trees

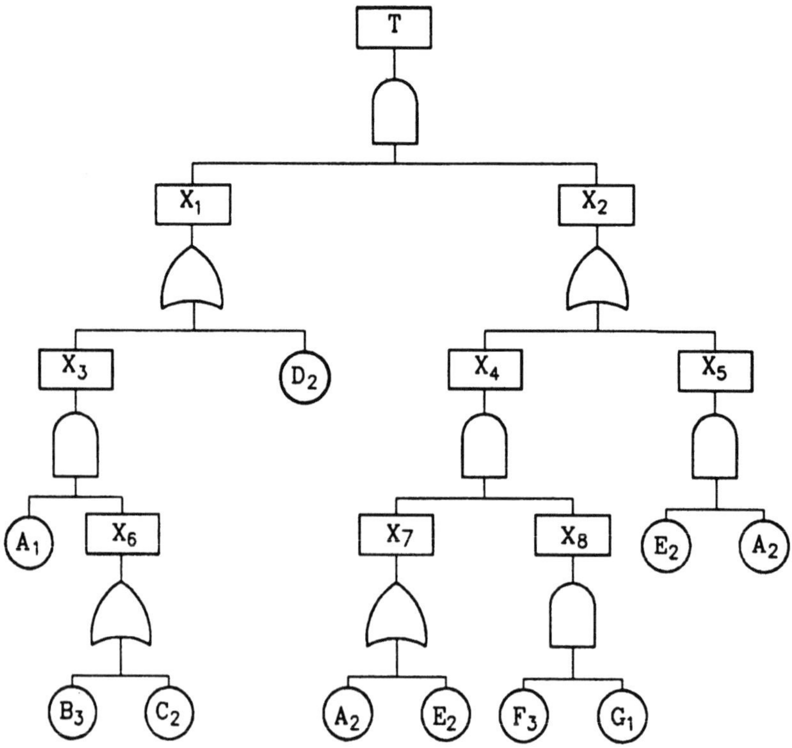

Fig. 9.1 Typical fault tree.

Table 9.1 Fault-tree symbols

Event symbols		*Gate symbols*		*Transfer symbols*	
 Circle	Basic event with sufficient data	 AND gate	Output occurs if all input events occur simultaneously	 Transfer-in	A subtree is connected to this point.
 Diamond	Undeveloped event	 OR gate	Output event occurs if one or more of the input events occur.	 Transfer-out	This tree is connected to a supertree.
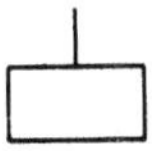 Rectangle	Gate output event	 EXOR gate	Output occurs when one, and not both, of the input events occur.		
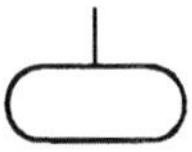 Oval	Conditional event	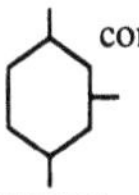 cond. INHIBIT gate	Conditional input event blocks the propagation of the input events.		
 House	Trigger event	 INV gate	Output event is true if and only if input event is false.		

representing events and their relations have been more or less standardized and can be seen in Table 9.1. Three types of symbols can be distinguished: event symbols, gate symbols and transfer symbols. The following event symbols are commonly used.

1. A circle represents a basic event such as the failure of an elementary component for which sufficient probability information is available and further development is not required.
2. A diamond expresses an undeveloped or incomplete event which is not developed further due to lack of time or interest, or because more detailed information is not available.
3. A rectangle signifies a gate output event or resulting event which results from the combination of input events through the corresponding gate. The top event itself is a resulting event as well.

4. An oval symbol denotes a conditional event. It indicates any condition or restriction that applies to a logic gate.
5. The house-shaped symbol indicates a trigger event, i.e. an event which is normally expected (or simulated to occur).

Gate symbols connect events according to their causal relations. The most important gates are as follows.

1. The AND gate indicates that an output event occurs if and only if all the input events occur simultaneously. The logic function represented by this gate is

$$Y_0 = A_a \cap B_b \cap C_c \cap \ldots \cap N_n. \tag{9.1}$$

2. The OR gate denotes that each input event is able to produce the output event irrespective of the other events. It describes the logic function

$$Y_0 = A_a \cup B_b \cup C_c \cup \ldots \cup N_n. \tag{9.2}$$

3. The Exclusive-OR gate or EXOR gate produces the output event if one and only one of the two input events applies. The output event can be expressed as a logic function of the input events as

$$Y_0 = (A_a \cap \bar{B}_b) \cup (\bar{A}_a \cap B_b). \tag{9.3}$$

   Exclusive-OR gates with more inputs produce output when an odd number of input events is true.
4. The conditional event of the inhibit gate when true, forbids the input event to propagate through this gate. It is a graphical representation of the logic function

$$Y_0 = A_a \cap \bar{B}_b. \tag{9.4}$$

5. The invertor-gate (INV) produces output if and only if the input event is absent.

Transfer symbols (denoted by a triangle) are used either to prevent repetition of identical parts of a fault tree or to connect parts of a fault tree which are spread over different pages. The transfer-in symbol denotes the existence of a subtree originating at this point. The transfer-out symbol indicates that the subtree descending from this point is an input to another fault tree (Fig. 9.2).

9.2 MANUAL FAULT-TREE CONSTRUCTION

The major objective of fault-tree construction is to visualize the causal relationship between a system failure mode and the basic events responsible. However, showing merely basic events and their final impact (system

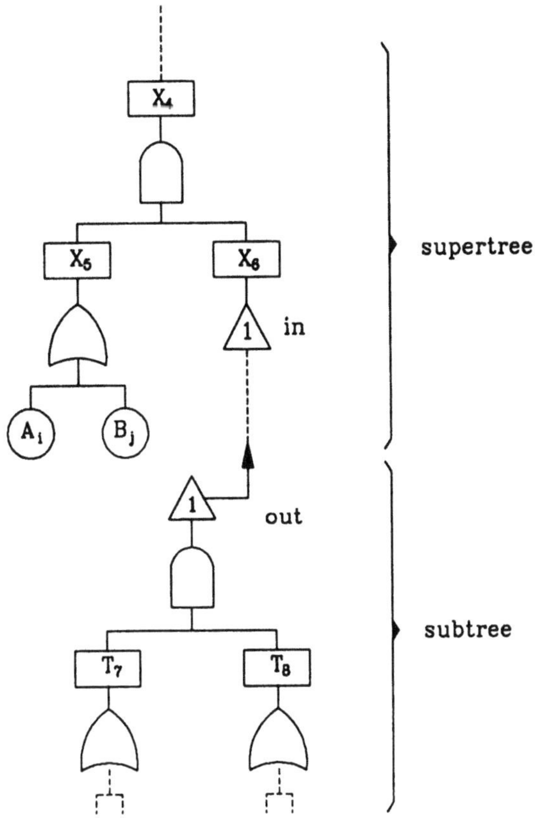

Fig. 9.2 Transfer-in and transfer-out symbols are used to connect subtrees with their supertrees.

failure) is not enough. A fault tree should also explain the way that the faults propagate through the system by showing (at least some) intermediate events. Only then the fault tree will be an appropriate tool to communicate with other parties, like clients and authorities who ask more and more often for complete fault-tree analyses to accompany product design and applications for licenses. Since most failures propagate through a system by the (normal) physical connections between subsystems (components) one often can gratefully make use of system descriptions like block diagrams made during system development. However, one should always be aware that these diagrams describe normal system behaviour in which some possible failures are omitted. In order to illustrate manual fault tree construction consider the following example.

EXAMPLE 9.1

A computer system for process control consist of two redundant computers A and B, each of which is able to control the process (Fig. 9.3). In

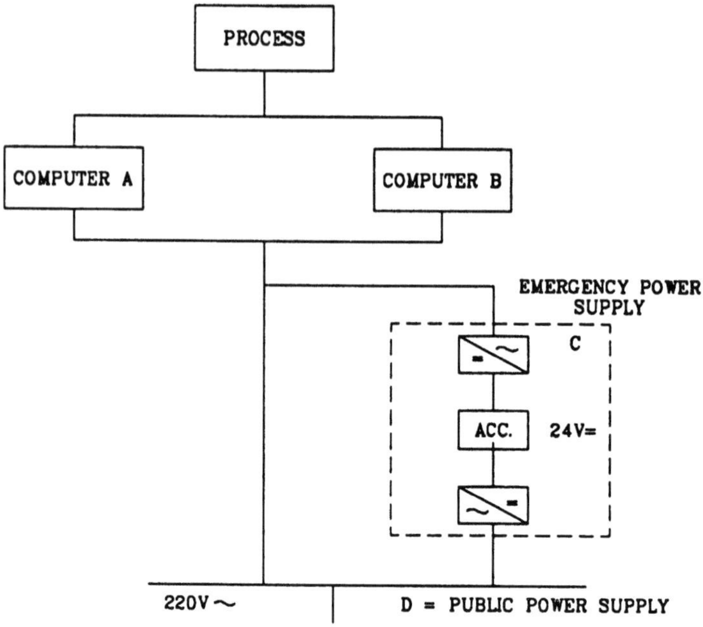

Fig. 9.3 Example system: a redundant computer system for process control.

order to overcome short power failures an emergency power supply C is added which is able to maintain the power supply for 1 hour until its accumulator is discharged completely. If the public power supply is not recovered within this hour the process will be uncontrolled.

SOLUTION

In this system the following independent component failures may occur:

1. computer A fails;
2. computer B fails;
3. emergency power supply C fails;
4. although it represents no component failure the discharged state of the emergency power supply does affect the system behaviour and so must be considered as well;
5. failure of public power supply D;
6. repair facility (E) fails to repair public power supply within 1 hour.

Now, let's consider the system failure event 'no process control'. From the system description it appears that this will occur when both computers show incorrect output which is visualized by the first AND gate in the fault tree of Fig. 9.4. Following down the block diagram it can be easily seen that the intermediate event X_1 (no control action by computer A) occurs when either one or both of the events $\bar{A}$ (failure of computer A) or X_{3a} (no power supply to computer A) occur, which is represented by an OR gate in the fault tree. Intermediate event X_2 can be developed in an identical manner. Since both events X_{3a} and X_{3b} obviously represent the same physical event X_3 (no power supply to computers) identical subtrees will descend from both points, which is indicated by the transfer-in and transfer-out symbols. Subsequently, intermediate event X_3 can be split up into two subevents 'public power supply failure' ($\bar{D}$) and 'no power supply from EPS' (X_4) by means of an AND gate.

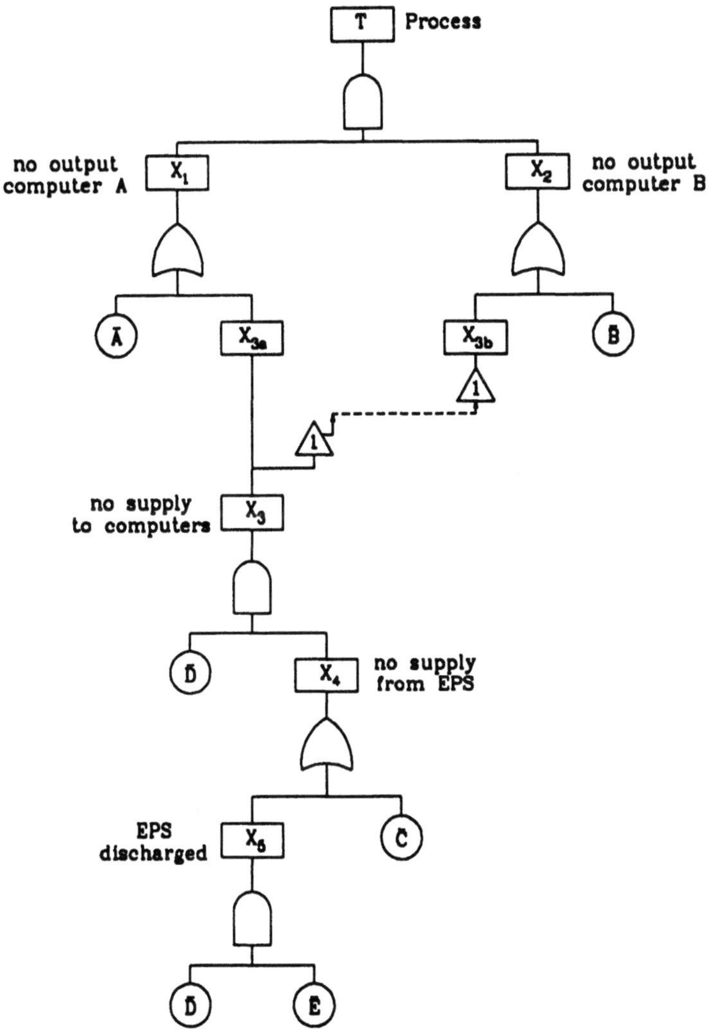

Fig. 9.4 Fault tree for top event 'no process control'.

Although the above procedure seems rather complicated for this simple system it will prove its value when applied to more complex system descriptions (e.g. when the subsystems of this example are split up into more basic components), since it minimizes the chances of overlooking basic events.

In the above example the system was supposed to contain two-state components only (although the intermediate event 'EPS discharged' can be regarded as a second failure mode of the emergency power supply). When multi-state components have to be considered the fault tree construction procedure does not change significantly. There is only a slight difference in notation: where the functional modes of a two-state component can be expressed as A and $\bar{A}$ the modes of a multi-state component must be notated as A_0 through A_n for a $n + 1$ state component. As an example consider the following system.

EXAMPLE 9.2

A safety system consists of three monitors A, B and C which all guard the same production process. Each monitor has two distinct failure modes:

1. the monitor generates a false alarm;
2. the monitor is unable to detect an unsafe situation.

The process should be switched off whenever a dangerous situation occurs. However, when the process is switched off unnecessarily due to a false alarm a significant loss of production will result. On the other hand when the safety system fails to switch off the process in a dangerous situation the production system might be damaged very badly. For reasons of reliability the system is switched off when at least two monitors show an unsafe situation. Figure 9.5 shows the fault tree for the event 'safety system fails' which combines both failures 'false alarm' and 'no detection'.

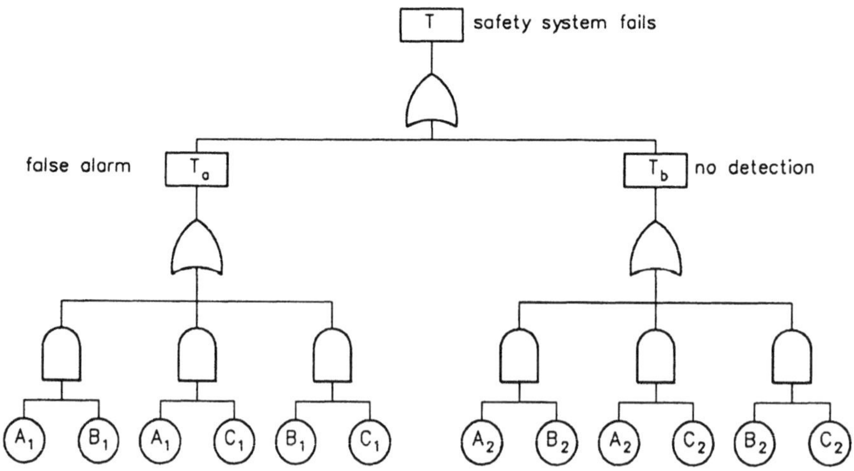

Fig. 9.5 Fault tree of the safety system of Example 9.2.

Success trees describing the successful action of a system can be constructed in a similar manner. However, when considering systems having multiple modes of operation the distinction between fault trees and success trees is less evident, since some (deteriorated) states might exist in which the system neither runs perfectly nor has failed completely. In the special case that a fault tree describes the malfunctioning of a system having only two-state components and comprises only AND, OR and INV gates the system's success tree can be extracted from the fault tree simply by exchanging AND and OR gates and replacing all events by their complementary events. INV gates are not affected by this procedure. Figure 9.6 shows the success tree of the redundant computer system depicted in Fig. 9.3 and is extracted from the fault tree in Fig. 9.4.

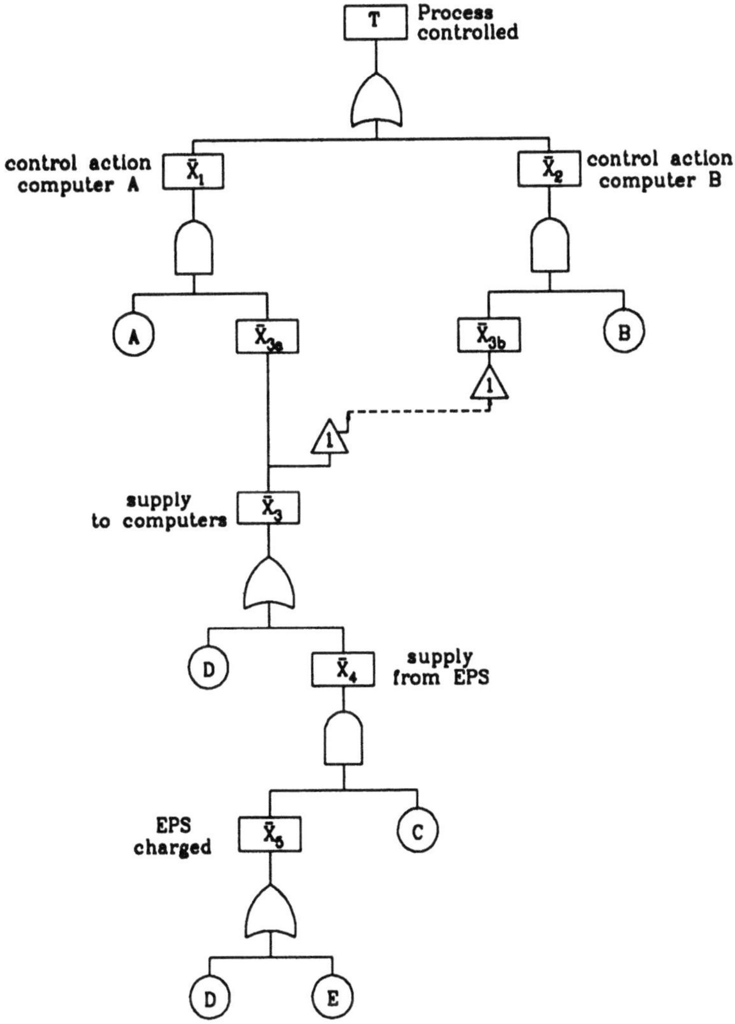

Fig. 9.6 Success tree for the redundant computer system of Example 9.1.

9.3 PRIME IMPLICANTS AND MINIMAL CUTS

In fault-tree analysis one is interested in all intersections (combinations) of basic events that lead to the occurrence of the undesired top event. Such an intersection is called an *implicant* because the simultaneous occurrence of the corresponding basic events implies the occurrence of the top event. One reason of interest is the insight into system behaviour that we gain when we know which combinations of basic faults result in system malfunction. A second reason is that a complete set of intersections forms a good basis for top probability calculations as will be shown in section 9.4. For large systems the number of fatal combinations is enormous since this number increases exponentially with the number of basic events.

However, most implicants are abundant in that they comprise more basic events than are required for system failure. Therefore, a restriction is applied, leaving only prime implicants that are minimal but sufficient conditions for the top event to take place. If one of the basic events is removed from a prime implicant the remaining set of events is no longer an implicant. Each prime implicant denotes a set of item states which is responsible for the occurrence of the top event. Hence the top event can be expressed as the union of all prime implicants. This logical function can be written as

$$T = P_1 \cup P_2 \cup P_3 \cup \ldots \cup P_n, \tag{9.5}$$

where P_i denotes the ith prime implicant. To ensure a complete description of the top event no prime implicant may be omitted. An efficient algorithm for obtaining prime implicants was introduced by Fussell and Vesely (1972). This algorithm is based on the feature that OR gates increase the number of prime implicants, whereas AND gates enlarge the size of prime implicants.

The algorithm starts with the implicant formed by the top event. This resultant event is expressed in terms of the input events to the corresponding gate. If this gate is an AND gate the resultant event in the implicant is replaced by the row of input events to this gate. In the case of an OR gate the implicant is duplicated once for each of the gate's input events, in which the resultant event is replaced by the input event under consideration. This process is repeated until the implicants contain no more resultant events. Implicants found to contain mutually exclusive events are inconsistent and must be deleted since they describe system states that are physically impossible. Finally, the set of prime implicants is found when all implicants not being prime implicants are eliminated. As an example the Fussell–Vesely algorithm has been applied to the fault tree of Fig. 9.4. The iterative process involved in shown in Fig. 9.7. One drawback of this algorithm is that it considers only AND and OR gates. This means that in the fault-tree synthesis process the use of other gates should be avoided, which is possible when fault propagation through a system

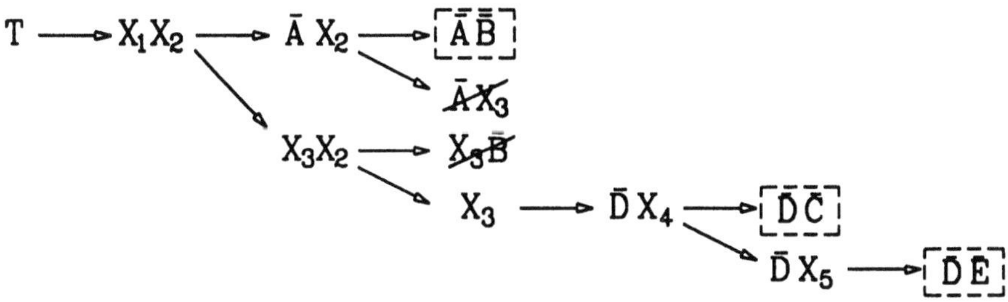

Fig. 9.7 Iterative process for separating out prime implicants from the fault tree depicted in Fig. 9.4. The prime implicants are indicated by dashed lines.

depends exclusively on the simultaneous occurrence of basic events and not on their order of appearance. Then the corresponding logic functions can be described using only AND gates, OR gates and complementary events. Exclusive-OR gates, for example, are very similar to OR gates except that they exclude the simultaneous occurrence of the input events. When the input events of an EXOR gate are both independent and rare, the probability of their simultaneous occurrence will be negligible in comparison with the probabilities of the single events. In these cases the EXOR gate might be replaced by a single OR gate without spoiling the credibility of the fault-tree analysis. Otherwise, the EXOR gate must be replaced by the combination of AND and OR gates shown in Fig. 9.8(a). Likewise, inhibit gates might be avoided by the construction of Fig. 9.8(b).

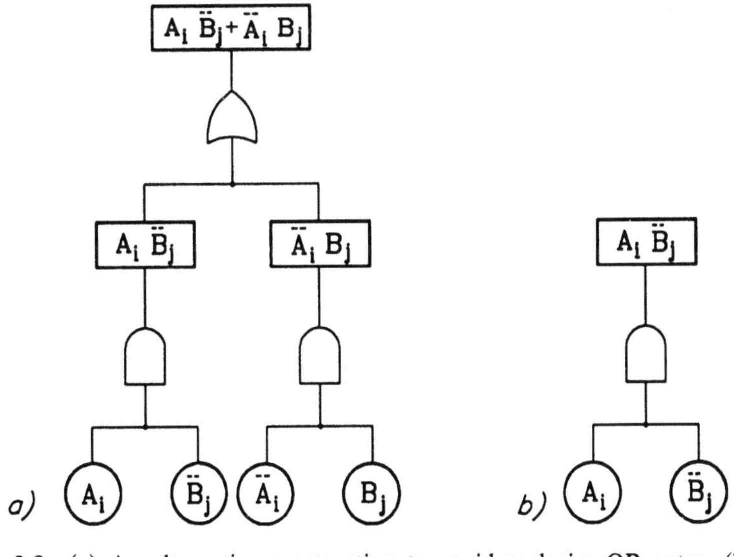

Fig. 9.8 (a) An alternative construction to avoid exclusive-OR gates. (b) An alternative notation to avoid inhibit gates.

As indicated above, the Fussell–Vesely algorithm is not capable of processing invertor gates which, regretfully, cannot always be avoided in a fault tree construction process. Fortunately, INV gates can be removed easily from fault trees. This elimination process proceeds as follows:

1. search the fault tree for INV gates from top to bottom.
2. Whenever an INV gate is found replace it by a direct throughput and modify the descending subtree in the following manner:
   (a) Exchange AND gates with OR gates and *vice versa* (INV gates are not affected).

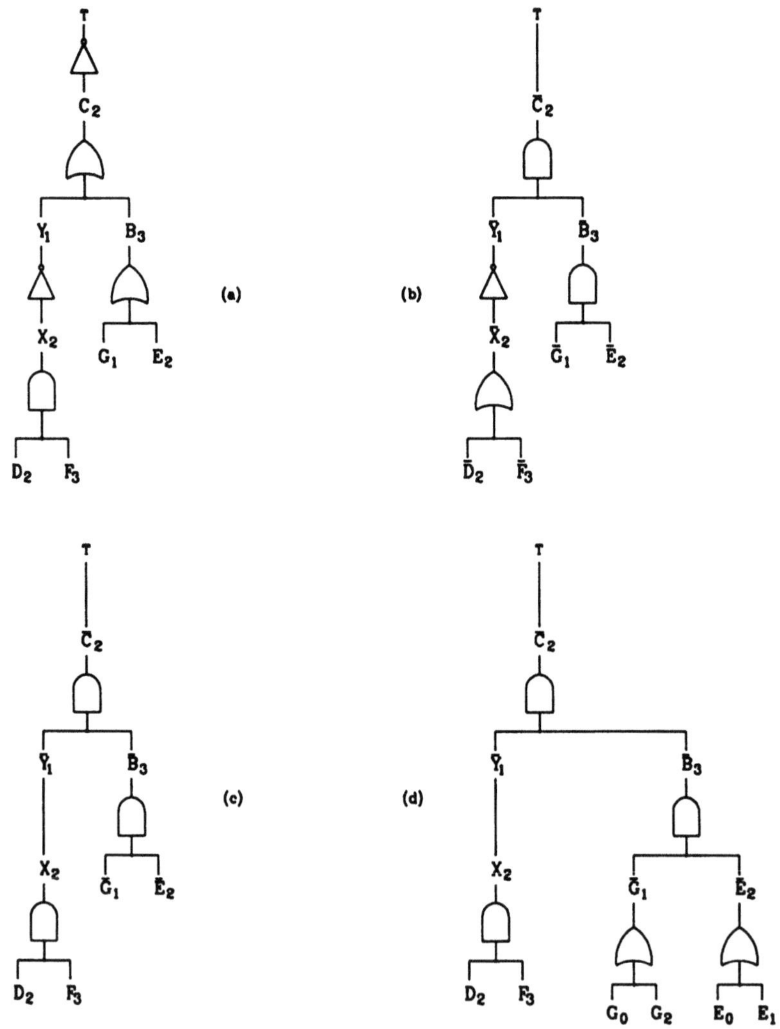

Fig. 9.9 Elimination of INV gates from a fault tree: (a) example tree having two INV gates; (b) first INV gate elimination; (c) second INV gate elimination; (d) additional OR gates to replace complementary events.

(b) Replace all events with their complementary events by adding negation marks or eliminating previously defined negation marks.

3. When all INV gates have been removed each complementary basic event will be made a resultant event by adding an OR gate having inputs for all disjunct states of the corresponding item. This elimination procedure is visualized in Fig. 9.9.

9.4 TOP PROBABILITY CALCULATION

One method of evaluating the top probability from basic event probabilities is to calculate the output event probabilities of the gates immediately above the fault tree's input events using the basic probability laws for logic intersection and logic union

$$Pr(A_i \cap B_j) = Pr(A_i) \cdot Pr(B_j | A_i) \tag{9.6}$$

and

$$Pr(A_i \cup B_j) = Pr(A_i) + Pr(B_j) - Pr(A_i \cap B_j). \tag{9.7}$$

In turn, these resulting probabilities can be used to calculate the output events of the next gates. This process might proceed until the top event is reached, which is illustrated by the example tree shown in Fig. 9.10. Here

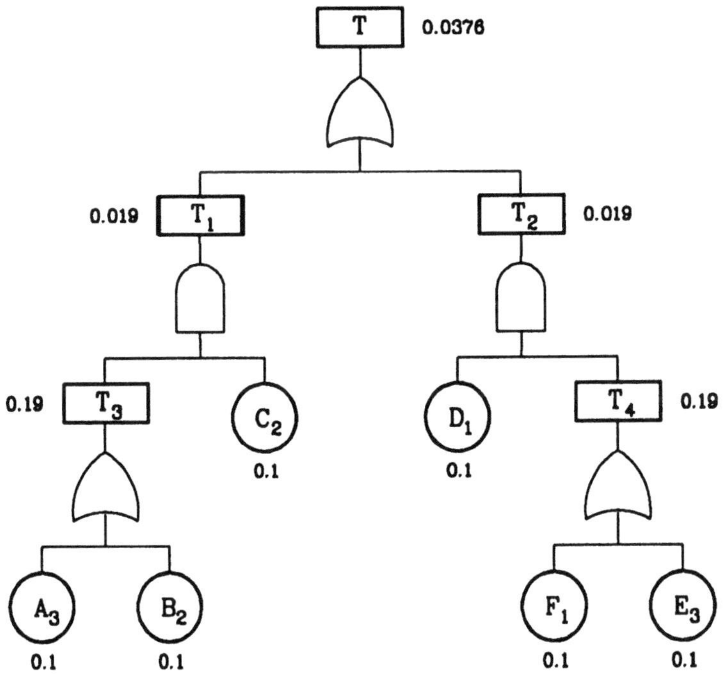

Fig. 9.10 Gate-by-gate enumeration of the top probability. The basic events are assumed to have a 0.1 probability.

all basic events are supposed to have a 0.1 probability of occurrence. However, once the resultant event probability of a gate is calculated all other information about the basic events responsible, especially information concerning dependencies, is lost. Hence, this method is applicable only to fault trees having no dependent basic events. This brings us to a major problem in top-event probability calculation, namely taking account of dependent events. Fault trees representing structure functions of univocal systems can be described by only AND and OR logic. However, they may still contain three types of statistically dependent events:

1. repeated events, i.e. events appearing more than once in a single fault tree;
2. mutually exclusive events, e.g. different states of the same component;
3. partially dependent events.

The first two dependencies are easily taken account of in the probability calculation process. The third type, however, creates a problem and should therefore be eliminated. All partially dependent events are composed of more elementary subevents which are either independent, repeated or mutually exclusive events. If one is able to extract those subevents and their probabilities for each partially dependent event a probability evaluation of the top event is possible using the following procedure which has a more general validity. First the top event is expressed as the union of all prime implicants

$$T = P_1 \cup P_2 \cup \ldots, \cup P_n. \tag{9.8}$$

Since each prime implicant is represented as an intersection of basic events no information concerning dependent events is lost. Using the inclusion-exclusion principle the top event's probability can be calculated as

$$\begin{aligned} Pr\{T\} &= \sum_{i=1}^{n} Pr\{P_i\} \\ &\quad - \sum_{\substack{i=2 \\ i \neq j}}^{n-1} Pr\{P_i \cap P_j\} \\ &\quad \vdots \\ &= + (-1)^{n+1} Pr\{P_1 \cap P_2 \cap \ldots \cap P_n\} \end{aligned} \tag{9.9}$$

Prior to the evaluation all terms in brackets should be checked for dependencies. When a term contains repeated basic events their redundant appearance must be eliminated. Terms comprising mutually exclusive basic events have a zero contribution to the top-event probability and should therefore be disregarded. After this elimination process only terms containing exclusively independent basic events are left; hence, their

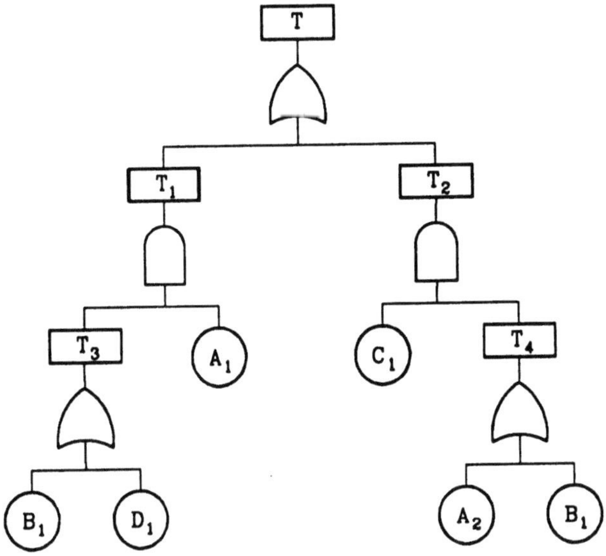

Fig. 9.11 A fault tree having both repeated and mutually exclusive basic events.

probabilities are simply the products of the basic event probabilities involved. As an example consider the fault tree depicted in Fig. 9.11, which has both repeated events (B_1) and mutually exclusive events (A_1, A_2). The top event in terms of prime implicants can be written as

$$T = A_1B_1 \cup A_1D_1 \cup A_2C_1 \cup B_1C_1. \tag{9.10}$$

Then the corresponding probability is given by:

$$\begin{aligned} Pr\{T\} &= Pr\{A_1B_1\} \cup Pr\{A_1D_1\} \cup Pr\{A_2C_1\} \cup Pr\{B_1C_1\} \\ &\quad - Pr\{A_1B_1D_1\} - Pr\{A_1B_1C_1\} - Pr\{A_1B_1C_1D_1\} \\ &\quad - Pr\{A_2B_1C_1\} + Pr\{A_1B_1C_1D_1\} \\ &= (a_1b_1 + a_1d_1 + a_2c_1 + b_1c_1) \\ &\quad - (a_1b_1d_1 + a_1b_1c_1 + a_1b_1c_1d_1 + a_2b_1c_1) + (a_1b_1c_1d_1) \\ &= \omega_1 - \omega_2 + \omega_3. \end{aligned} \tag{9.11}$$

This latter probability evaluation method is not obstructed by mutually exclusive or repeated events and is therefore well suited to analyse systems having multi-state components. The exact evaluation of the top-event's probability requires the enumeration of all possible terms. However, the number of combinations increases excessively with the number of prime implicants. Approximated values can be obtained when the evaluation is stopped after the kth order combinations. The approximation is an upper bound for the top-event's probability when k is odd, while lower bounds are obtained when k is even. Then the maximum error involved with the k-th order approximation can be expressed as: $|\varepsilon_k| \leqslant \omega_k$.

9.5 TOP FREQUENCY CALCULATION

Top probability evaluation alone is insufficient to describe the system's stochastic behaviour since it does not specify the rate at which failures can be expected. Therefore, failure frequency evaluation is required as well to complete the reliability analysis. Several algorithms for calculating the frequency with which the top event of a fault tree occurs have been discussed in the literature (see the bibliography to Chapter 9). However, all these algorithms are confined to systems having only two-state components. In this section two algorithms, the top event's frequency algorithm (Bossche, 1984) and the system interstate frequency algorithm (Bossche, 1986) are introduced. The first algorithm is able to evaluate the exit frequency of a fault tree's top event, i.e. the frequency $f(\bar{T})$ that the system leaves the state that corresponds to the top event. The latter algorithm (SIFA) can produce transition frequencies f_{ij} between any pair of disjunct system states (i,j) which are not necessarily each other's complement as was the case in the TEFA algorithm. This means that it can be used in a recursive manner, i.e. when it is used to evaluate the transition frequencies between *all* disjoint states of a subsystem a component representation of the subsystem is available and can be used as input to fault trees of larger (sub)systems. Both algorithms can be applied to systems with multistate components and will be introduced in this section.

9.5.1 The TEFA algorithm

Multi-state components may have a lot of interstate transitions. For example, Fig. 9.12 shows all possible transitions for a three-state component. Each transition of a basic component may cause the system to transfer out its current state. When a logic expression of the system state, such as a fault tree, is available, the exit frequency of this state can be evaluated by a single application of the inclusion-exclusion rule. In the top-event's frequency algorithm (TEFA) the contributions of all intersections of prime implicants are calculated and summed in order to obtain the top event's exit frequency. Then this frequency can be expressed as

$$f_{\bar{T}} = \sum_{k=1}^{n} (-1)^{k+1} \nu_k, \tag{9.12}$$

where ν_k is the joint contribution of all kth-order combinations of prime implicants.

Three types of combinations, each having a different contribution, can be distinguished:

1. A combination of prime implicants which contains no disjunctions (mutually exclusive basic events), e.g. $P_2 \ldots, Z_0$, contributes to the

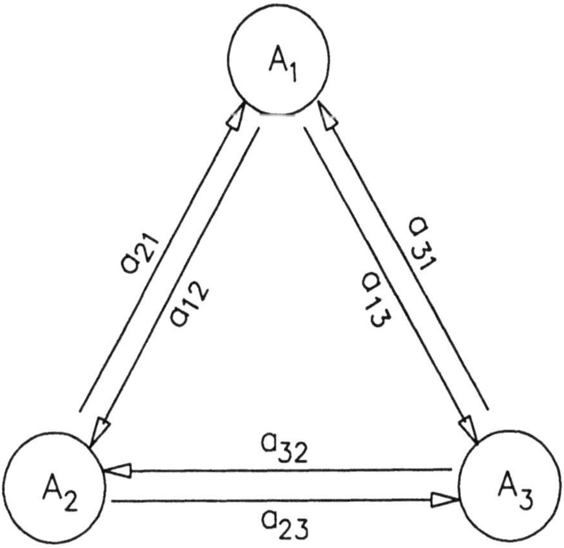

Fig. 9.12 Possible transitions for a three-state component.

top frequency through all transitions that abolish one of the combination's basic events. In a mathematical formula this (signed) contribution can be written as

$$(-1)^{k+1} Pr\{P_2\} \ldots Pr\{Z_0\} \left[\sum_{j \neq 2} \mu(p_{2j}) + \ldots + \sum_{j \neq 0} \mu(z_{0j}) \right], \quad (9.13)$$

where k is the number of prime implicants used for the combination.

2. A combination that contains exactly one disjunction, e.g. $A_1A_2P_2 \ldots$, Z_0, has a contribution to the top frequency which is based upon only those transitions that bridge the disjunction. The descriptive formula is then

$$(-1)^{k+1} Pr\{P_2\} \ldots Pr\{Z_0\}[Pr\{A_1\}\mu(a_{12}) + Pr\{A_2\}\mu(a_{21})]. \quad (9.14)$$

3. For all prime implicant combinations containing more than one disjunction, e.g. $A_1A_2A_3 \ldots$, Z_1 or $A_1A_2P_1 \ldots$, $R_2R_3 \ldots$, Z_2, the contribution equals 0.

Sometimes the top event is not described by a set of prime implicants but by a set of mutually disjoint implicants. In that case the above algorithm will attain the exact frequency of the top event after the second-order combinations of implicants because the higher-order combinations contain at least two disjunctions. To calculate the steady-state frequency of a system failure, the TEFA algorithm can be applied to the logic expression for this system failure as well as that for the complementary event, because in the steady-state case the frequency of entering a system state equals the frequency of leaving this stage. For non-steady-state situations the logic expression for the complementary event $\bar{T}$ should be used to evaluate the frequency of entering the system state described by the top event T, while the expression for the top event is required to calculate the frequency of leaving the corresponding system state.

Again, approximated values can be obtained when the evaluation is stopped after the kth-order combinations. The approximation is an upper bound for the top-event's frequency when k is odd, while lower bounds are obtained when k is even. The maximum error involved with the kth-order approximation (ε_k) obeys the following relation

$$|\varepsilon_k| \leqslant \nu_k. \tag{9.15}$$

Since transition rates and probabilities are defined as positive all ν_k are positive as well.

EXAMPLE 9.3

As an example consider two electric power units A and B, each of which can supply 75% of the power required. The units have three states: a working state, a degraded state and a failed state. Degraded units work at half power. This means that the power-supply system fails when one of the units is failed or when both units are degraded. Figure 9.13 shows the fault tree for this system while Table 9.2 shows all the terms found by the inclusion-exclusion rule, and their respective contributions to the top-event's probability and frequency. Because the contributions of the third-order term equal 0, exact figures for the top-event's steady state probability and frequency are obtained by a second-order approximation. As can easily be seen from Table 9.2 the first-order approximation yields upper bounds for both quantities.

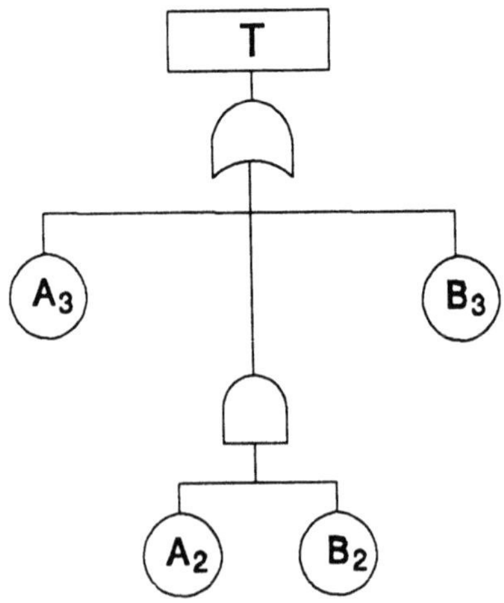

Fig. 9.13 Fault tree for the two-unit power-supply system of Example 9.3.

Table 9.2 Terms found by the inclusion-exclusion rule and their contributions to the top-event's probability and frequency

Term	*Contribution to top-event's* Probability	Frequency
A_3	a_3	$a_3\{\mu(a_{31}) + \mu(a_{32})\}$
B_3	b_3	$b_3\{\mu(b_{31}) + \mu(b_{32})\}$
A_2B_2	a_2b_2	$a_2b_2\{\mu(a_{21}) + \mu(a_{23}) + \mu(b_{21}) + \mu(b_{23})\}$
A_3B_3	a_3b_3	$-a_3b_3\{\mu(a_{31}) + \mu(a_{32}) + \mu(b_{31}) + \mu(b_{32})\}$
$A_2B_2B_3$	0	$-a_2\{b_2\mu(b_{23}) + b_3\mu(b_{32})\}$
$A_2A_3B_2$	0	$-b_2\{a_2\mu(a_{23}) + \alpha_3\mu(a_{32})\}$
$A_2A_3B_2B_3$	0	0

9.5.2 The SIFA algortihm

In order to limit the computational effort involved with reliability analysis, complex systems are often split up into subsystems, each realizing a clearly defined subfunction. Then the reliability analysis can be performed in a recursive manner, i.e. the probabilities and frequencies of subsystem failures are calculated prior to the evaluation of the probability and frequency data of the system. However, the algorithm for the calculation of the top-event's frequency introduced above produces output data (state frequencies) that are dissimilar to its input data (transition rates); hence, this algorithm is not appropriate for a recursive frequency analysis of complex systems. Therefore, a second algorithm, the system interstate frequency algorithm (SIFA), will be introduced which is able to evaluate the interstate frequencies of a (sub)system from which the system's transition rates can easily be found. In this algorithm the contributions of all intersection of at least one prime implicant of Φ_i and at least one prime implicant of Φ_j are calculated and summed. It offers the possibility of calculating the contributions of all component transitions to both the system transition ϕ_{ij} and the system transition ϕ_{ji} in one single search. Two principally different combinations of prime implicants can be distinguished:

1. A combination may contain exactly one disjunction, e.g. $A_pA_qR_r \dots Z_z$, where A_p is contributed by Φ_i, while A_q descends from Φ_j. The contributions of such a combination are
$$\Delta f(\phi_{ij}) = (-1)^{v+w}\mu(a_{pq})Pr\{A_p\}Pr\{R_r \dots Z_z\} \qquad (9.16)$$
to the interstate frequency $f(\phi_{ij})$ and
$$\Delta f(\phi_{ji}) = (-1)^{v+w}\mu(a_{qp})Pr\{A_q\}Pr\{R_r \dots Z_z\} \qquad (9.17)$$
to $f(\phi_{ji})$, where v and w represent the numbers of prime implicants used from the sets ϕ_i and ϕ_j, respectively.
2. For all combinations containing more than one disjunction the contributions equal zero.

Adding all contributions yields the interstate frequencies $f(\phi_{ij})$ and $f(\phi_{ji})$. Then the corresponding transition rates can be found using the relation:

$$\mu(\phi_{ij}) = \frac{f(\phi_{ij})}{Pr\{\phi_i\}}. \tag{9.18}$$

These rates can, in turn, be used as input to the algorithm described above. When all interstate transition frequencies of the subsystem have been calculated it is possible to derive the state frequencies, i.e. the frequency of entering the system state, using the formula:

$$f(\phi_i) = \sum_{\substack{j=1 \\ j \neq i}}^{n} f(\phi_{ji}) \tag{9.19}$$

while the frequency of leaving this state can be calculated as

$$f(\bar{\phi}_i) = \sum_{\substack{j=1 \\ j \neq i}}^{n} f(\phi_{ij}). \tag{9.20}$$

The SIFA algorithm introduced here is able to produce the interstate frequencies as well as the interstate transition rates of subsystems which, in turn, can be used as input to the same algorithm in order to calculate the system's interstate frequencies. A set of prime implicants is required for each state, including the success state, of the (sub)system. In the case of an n-state subsystem a maximum of $n(n - 1)/2$ computation have to be carried out. Hence, application of the algorithm involves a considerable amount of computational effort. On the other hand, some savings in time can be achieved by subdividing the system into modules.

EXAMPLE 9.4

A product is created by a production system in two steps. Both stages of the production line consist of two machines in parallel (Fig. 9.14(a)).

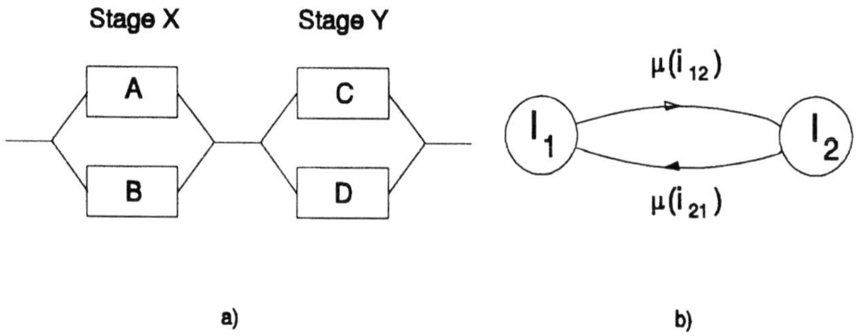

Fig. 9.14 Production system of Example 9.4: (a) schematic diagram; (b) state diagram of individual machines.

Table 9.3 The logic expressions, probabilities and interstate frequencies for the production system of Example 9.4

States	*Subsystem X* State probabilities	Interstate frequencies
$X_1 = A_1B_1$	$x_1 = a_1b_1$	$f(x_{12}) = a_1b_1\{\mu(a_{12}) + \mu(b_{12})\}$
$X_2 = A_1B_2 + A_2B_1$	$x_2 = a_1b_2 + a_2b_1\ (-0)$	$f(x_{21}) = a_1b_2\mu(b_{21}) + a_2b_1\mu(a_{21})$
$X_3 = A_2B_2$	$x_3 = a_2b_2$	$f(x_{13}) = f(x_{31}) = 0$
		$f(x_{23}) = a_1b_2\mu(a_{12}) + a_2b_1\mu(b_{12})$
		$f(x_{32}) = a_2b_2\{\mu(a_{21}) + \mu(b_{21})\}$
States	***Subsystem Y*** **State probabilities**	**Interstate frequencies**
$Y_1 = C_1D_1$	$y_1 = c_1d_1$	$f(y_{12}) = c_1d_1\{\mu(c_{12}) + \mu(d_{12})\}$
$Y_2 = C_1D_2 + C_2D_1$	$y_2 = c_1d_2 + c_2d_1\ (-0)$	$f(y_{21}) = c_1d_2\mu(d_{21}) + c_2d_1\mu(c_{21})$
$Y_3 = C_2D_2$	$y_3 = c_2d_2$	$f(y_{13}) = f(y_{31}) = 0$
		$f(y_{23}) = c_1d_2\mu(c_{12}) + c_2d_1\mu(d_{12})$
		$f(y_{32}) = c_2d_2\{\mu(c_{21}) + \mu(d_{21})\}$
States	***Main system*** **State probabilities**	**Interstate frequencies**
$\Phi_1 = X_1Y_1$	$\phi_1 = x_1y_1$	$f(\phi_{12}) = x_1y_1\{\mu(x_{12}) + \mu(y_{12})\}$
$\Phi_2 = X_1Y_2 + X_2Y_1 + X_2Y_2$	$\phi_2 = x_1y_2 + x_2y_1 + x_2y_2$	$f(\phi_{21}) = x_1y_2\mu(y_{21}) + x_2y_1\mu(x_{21})$
$\Phi_3 = X_3 + Y_3$	$\phi_3 = x_3 + y_3 - x_3y_3$	$f(\phi_{13}) = f(\phi_{31}) = 0$
		$f(\phi_{23}) = x_1y_2\mu(y_{23}) + x_2y_1\mu(x_{23})$ $x_2y_2\{\mu(x_{23}) + \mu(y_{23})\}$
		$f(\phi_{32}) = x_1y_3\mu(y_{32}) + x_3y_1\mu(x_{32})$ $x_2y_3\mu(y_{32}) + x_3y_2\mu(x_{32})$

The machines in stage X have a capacity of 60% of the desired production level each, while the machines in stage Y have 50% capacity each. All machines have independent ROCOFs $\mu(i_{12})$ and repair rates $\mu(i_{21})$ as can be seen in Fig. 9.14(b). From a production point of view three different system states Φ_1, Φ_2 and Φ_3 are defined corresponding to production levels >75%, 50–75% and <50%, respectively.

Table 9.3 shows the logic expressions, the probabilities and the interstate frequencies for these states.

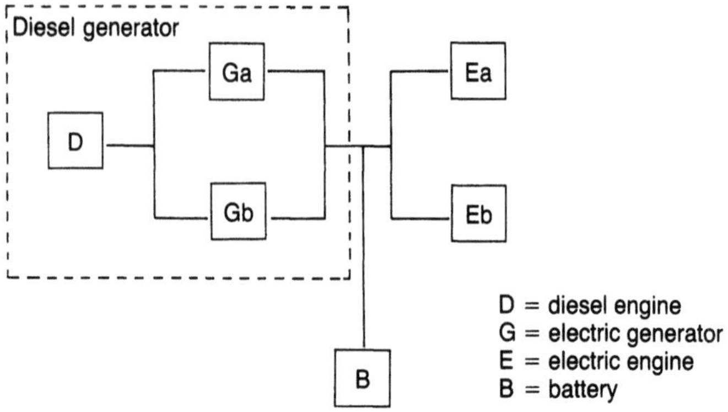

Fig. 9.15 Schematic diagram of the power-supply and propulsion system of the submarine in Problem 9.1.

9.6 WORKED PROBLEMS

PROBLEM 9.1

A submarine has two electric engines for its propulsion. The electric energy is generated by an diesel generator consisting of two electric generators coupled to a diesel engine (Fig. 9.15). On failure of the diesel generator the electric power supply is taken over by a battery which has sufficient energy to maintain cruise speed for five hours.

The battery may fail due to a short circuit in both active and standby mode. The two electric generators and the two electric engines are in active redundancy, i.e. one of each may fail without loss of the respective functions.

Further information:

1. repair of an item starts upon failure;
2. each item has its own repair facility;
3. since all items are complex all failure and repair times have exponential p.d.f.s by approximation. The effective failure and repair rates for each item are given in Table 9.4.

Table 9.4 Failure rates (ROCOFs) and repair rates for the components in Fig. 9.15

Item	λ	μ	*Unavailability*
Diesel engine	10^{-3}	4×10^{-1}	2.5×10^{-3}
Electrical generator	10^{-4}	2×10^{-1}	5×10^{-4}
Electrical engine	10^{-4}	2×10^{-1}	5×10^{-4}
Battery	10^{-6}	2×10^{-2}	5×10^{-5}

1. Set up the fault tree for the event 'no propulsion'.
2. Give the minimal cuts (prime implicants) for this event.
3. Evaluate the probability of the top event.
4. Evaluate the frequency of the top event.

ANSWER

1.

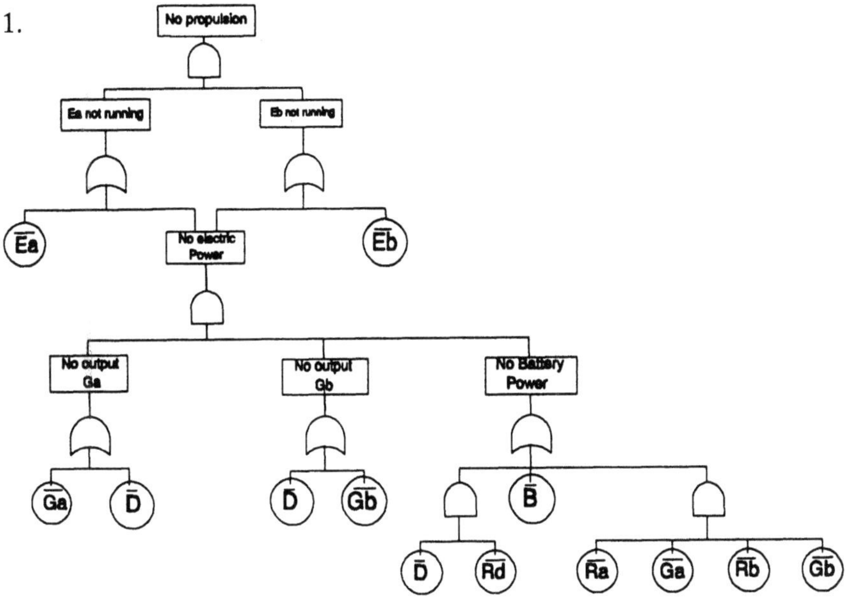

Fig. 9.16 Fault tree for Problem 9.1.

2. With help of the Fussel–Vesely algorithm the following minimal cuts can be found

$$\bar{D}\bar{R}_{\mathrm{d}} \quad \bar{D}\bar{B} \quad \bar{E}_{\mathrm{a}}\bar{E}_{\mathrm{b}} \quad \bar{G}_{\mathrm{a}}\bar{G}_{\mathrm{b}}\bar{B} \quad \bar{G}_{\mathrm{a}}\bar{G}_{\mathrm{b}}\bar{R}_{\mathrm{a}}\bar{R}_{\mathrm{b}},$$

where: $\bar{R}_{\mathrm{d}}$ = Diesel not repaired within five hours ($\bar{r}_{\mathrm{d}} = e^{-2}$); $\bar{R}_{\mathrm{a}}$, $\bar{R}_{\mathrm{b}}$ = Generator G_{a} (G_{b}) not repaired within five hours after second generator failure ($\bar{r}_{\mathrm{a}} = \bar{r}_{\mathrm{b}} = e^{-1}$).

3. The disjoint term method gives the following exact result:

$$\begin{aligned} Pr(T) &= \bar{d}\bar{r}_{\mathrm{d}} + \bar{d}\bar{b}r_{\mathrm{d}} + \bar{e}_{\mathrm{a}}\bar{e}_{\mathrm{b}}d + \bar{e}_{\mathrm{a}}\bar{e}_{\mathrm{b}}\bar{d}r_{\mathrm{d}}b + \bar{g}_{\mathrm{a}}\bar{g}_{\mathrm{b}}\bar{b}de_{\mathrm{a}} \\ &+ \bar{g}_{\mathrm{a}}\bar{g}_{\mathrm{b}}\bar{b}\bar{e}_{\mathrm{a}}de_{\mathrm{b}} + \bar{g}_{\mathrm{a}}\bar{g}_{\mathrm{b}}\bar{r}_{\mathrm{a}}\bar{r}_{\mathrm{b}}bde_{\mathrm{a}} + \bar{g}_{\mathrm{a}}\bar{g}_{\mathrm{b}}\bar{r}_{\mathrm{a}}\bar{r}_{\mathrm{b}}bd\bar{e}_{\mathrm{a}}e_{\mathrm{b}} \\ &+ \bar{g}_{\mathrm{a}}\bar{g}_{\mathrm{b}}\bar{r}_{\mathrm{a}}\bar{r}_{\mathrm{b}}b\bar{d}r_{\mathrm{d}}e_{\mathrm{a}} + \bar{g}_{\mathrm{a}}\bar{g}_{\mathrm{b}}\bar{r}_{\mathrm{a}}\bar{r}_{\mathrm{b}}b\bar{d}r_{\mathrm{d}}\bar{e}_{\mathrm{a}}e_{\mathrm{b}} \\ &\approx 3.3869 \times 10^{-4}. \end{aligned} \tag{9.21}$$

A first-order approximation with the inclusion-exclusion rule shows to be a very good estimate as well

$$Pr(T) \approx \bar{d}\bar{r}_{\mathrm{d}} + \bar{d}\bar{b} + \bar{e}_{\mathrm{a}}\bar{e}_{\mathrm{b}} + \bar{g}_{\mathrm{a}}\bar{g}_{\mathrm{b}}\bar{b} + \bar{g}_{\mathrm{a}}\bar{g}_{\mathrm{b}}\bar{r}_{\mathrm{a}}\bar{r}_{\mathrm{b}}$$
$$\approx 3.3875 \times 10^{-4}. \qquad (9.22)$$

4. The steady-state failure frequency can be approximated with a first-order application of the TEFA algorithm as

$$f_T \approx \bar{d}\bar{r}_{\mathrm{d}}(\mu_{\mathrm{d}}) + \bar{d}\bar{b}(\mu_{\mathrm{d}} + \mu_{\mathrm{b}}) + \bar{e}_{\mathrm{a}}\bar{e}_{\mathrm{b}}(2\mu_{\mathrm{e}})$$
$$+ \bar{g}_{\mathrm{a}}\bar{g}_{\mathrm{b}}\bar{b}(2\mu_{\mathrm{g}} + \mu_{\mathrm{b}}) + \bar{g}_{\mathrm{a}}\bar{g}_{\mathrm{b}}\bar{r}_{\mathrm{a}}\bar{r}_{\mathrm{b}}(2\mu_{\mathrm{g}}) \qquad (9.23)$$
$$\approx 1.355 \times 10^{-4}\,\mathrm{h}^{-1}$$

PROBLEM 9.2

Figure 9.17 shows a subsystem of a chemical plant used to cool the nitric acid (HNO_3) produced in the previous process step to such a temperature that it can be fed safely into the reactor of the next stage where the nitric acid reacts with benzene to form nitrobenzene. A too high temperature may cause a dangerous (explosive) situation in this next stage. To prevent this, the output temperature of the subsystem is controlled by a feedback loop. Independent of this loop an extra safety system is added consisting of a shut-down unit which cuts the nitric acid flow in case the output temperature is threatening to become too high after all.

The following failures may occur in the feedback loop:

A_1 The temperature sensor fails reporting a false low temperature.
B_1 The controller fails making a false signal to 'valve 2' to close down the coolwater flow.

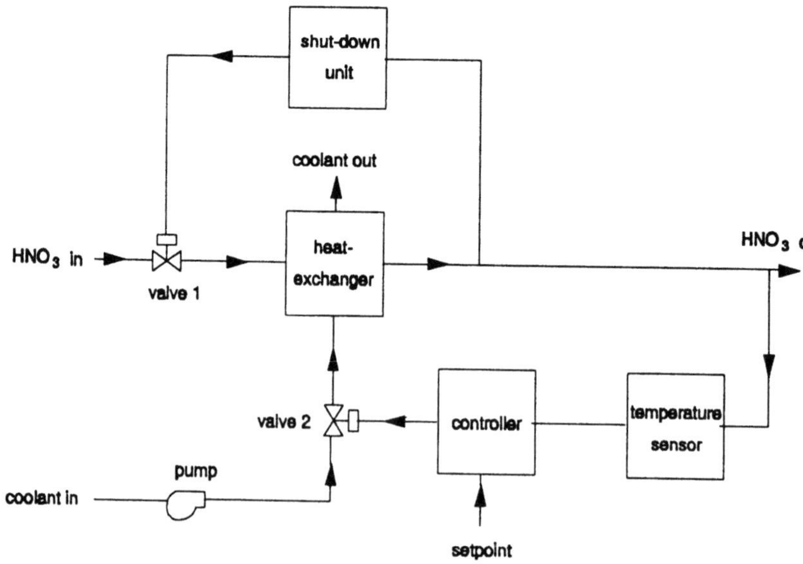

Fig. 9.17 Schematic diagram of the system in Problem 9.2.

C_1 The setpoint of the controller has been set to a too high temperature by a human error of the operator.

Other failures that may occur are:

D_1 The shutdown unit fails in such a mode that it cannot react to a high temperature signal.
E_1 Failure of the pump in the coolwater circuit.

All failure modes are assumed to occur statistically independent.

1. Construct a fault tree for the critical event Φ_1: 'output temperature too high'.
2. Extract the minimal cuts for this tree.
3. When all failures have an occurrence probability of q, evaluate the probability for the top event.
4. When each failure has a ROCOF of λ and a mean repair time of $1/\mu$ evalute the occurrence frequency of the top event.

ANSWER

1. The fault tree for this exercise is shown in Fig. 9.18.

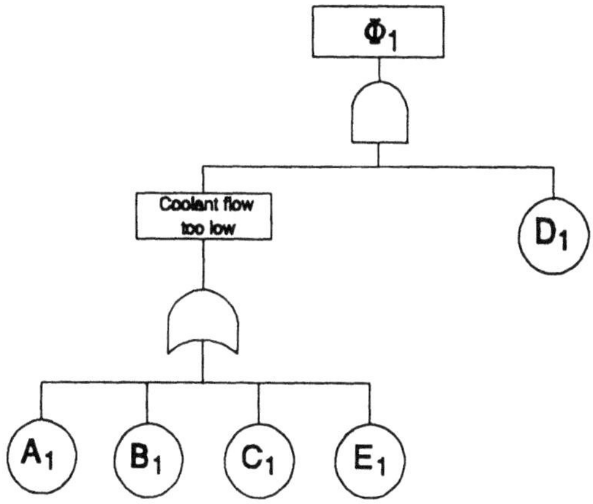

Fig. 9.18 Fault tree for Problem 9.2.

2. The minimal cut expression for this tree is

$$\Phi = A_1D_1 + B_1D_1 + C_1D_1 + E_1D_1. \tag{9.24}$$

3. The occurrence probabilities of the component failures are $q = \lambda/(\lambda + \mu)$, which result in a system failure probability of

$$\begin{aligned} Pr(\Phi_1) &= a_1d_1 + b_1d_1 + c_1d_1 + e_1d_1 - a_1b_1d_1 - a_1c_1d_1 \\ &\quad - a_1e_1d_1 - b_1c_1d_1 - b_1e_1d_1 - c_1e_1d_1 + a_1b_1c_1d_1 \\ &\quad + a_1b_1e_1d_1 + a_1c_1e_1d_1 + b_1c_1e_1d_1 - a_1b_1c_1e_1d_1 \\ &= 4q^2 - 6q^3 + 4q^4 - q^5. \end{aligned} \tag{9.25}$$

4. With help of the TEFA algorithm the frequency of the top event can be found to be

$$f_{\Phi_1} = 4q^2 2\mu - 6q^3 3\mu + 4q^4 4\mu - q^5 5\mu. \tag{9.26}$$

PROBLEM 9.3

A four-engined jet plane is able to continue its flight with only one engine operating on each side. Each engine obtains its fuel from the fuel tank on the corresponding wing by means of a fuel pump (Fig. 9.19).

The engines and pumps may fail while the fuel tanks may run dry.

1. Construct a fault tree for the event 'crash due to insufficient propulsion'
2. Extract the minimal cuts for this event.

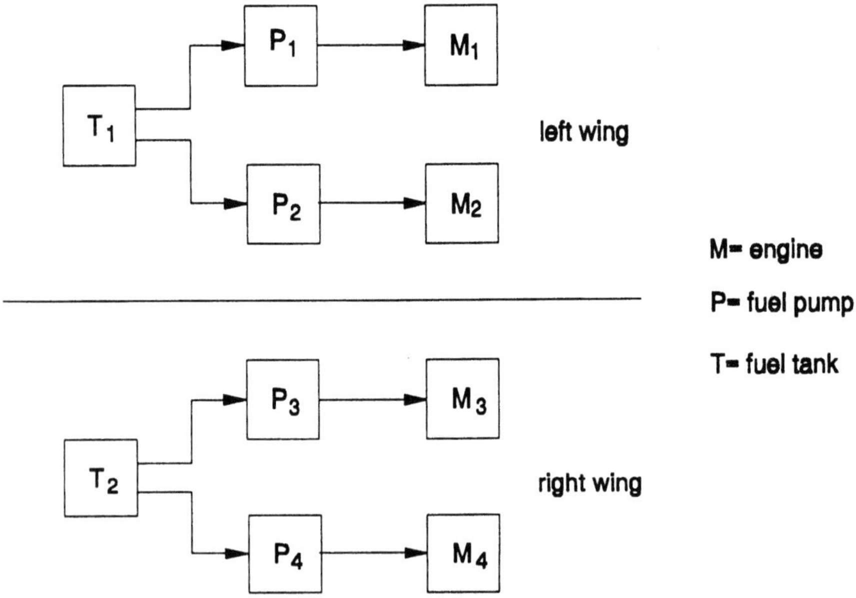

Fig. 9.19 Schematic diagram of the fuel and propulsion system of a four-engined jet.

ANSWER

1. The fault tree for this event is shown in Fig. 9.20.
2. The following minimal cuts can be found with help of the Fussell–Vesely algorithm:

$$\begin{array}{ccccc} T_1 & M_1M_2 & M_2P_1 & M_1P_2 & P_1P_2 \\ T_2 & M_3M_4 & M_4P_3 & M_3P_4 & P_3P_4 \end{array}$$

PROBLEM 9.4

Consider the system of Problem 9.2 again. In addition to the critical system failure due to too high an output temperature the system may also

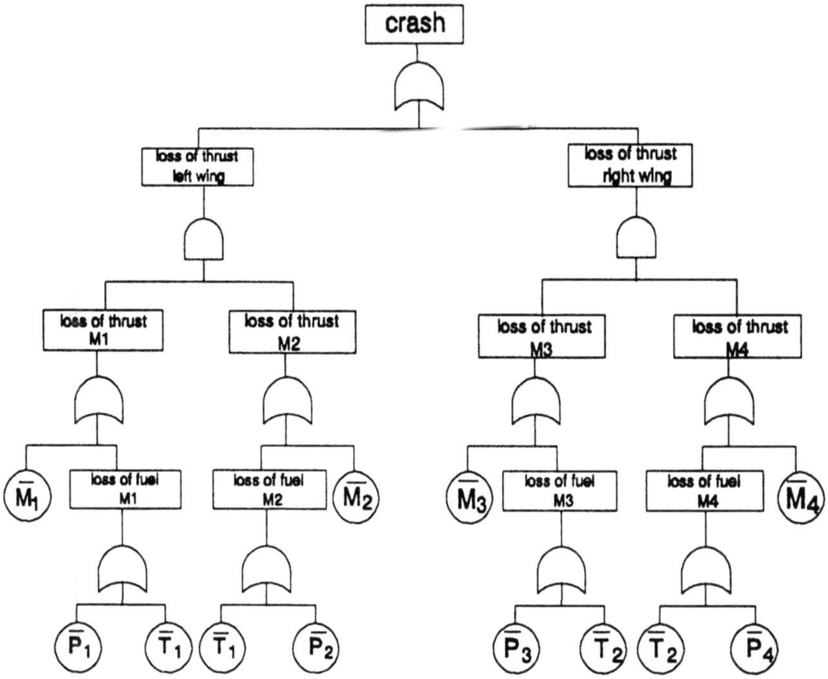

Fig. 9.20 Fault tree for Problem 9.3.

fail in a safe manner by producing a nitric acid flow of a too low temperature Φ_2 (due to one of the failures A_2 through C_2 below) or by producing no flow at all as a result of a successful shutdown (system state Φ_3).

The additional basic failures to be considered are:

A_2 The temperature sensor fails reporting a false high temperature.
B_2 The controller fails making a false signal to 'valve 2' to open completely.
C_2 The setpoint of the controller has been set to a too low temperature by a human error of the operator.

1. Find the expressions (in minimal cuts) for the events Φ_2 = 'output temperature too low' and Φ_3 = 'no output flow after shutdown'.

   The temperature sensor A, the controller B and the setpoint obey the three-state transition diagram depicted in Fig. 9.21(a), while the shutdown unit D and the cooling water pump follow the two-state transition diagram of Fig. 9.21(b).
2. Evaluate the failure frequencies of the three system failure modes.

ANSWER

1. The output temperature becomes too low either when failures in the feedback loop cause a too high coolant flow (Φ_2) or as a result of a

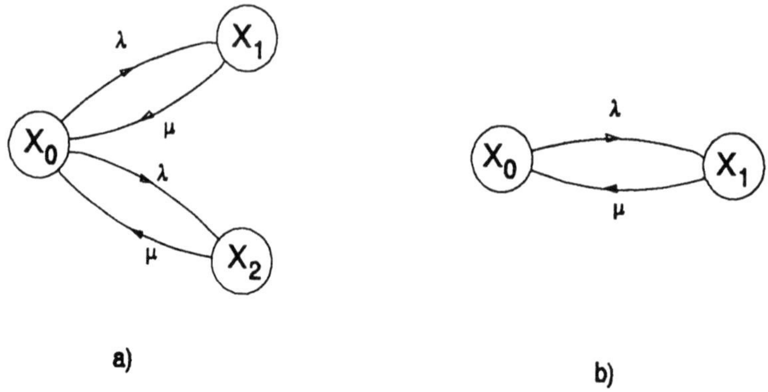

Fig. 9.21 State-space diagrams for the components in Problem 9.4.

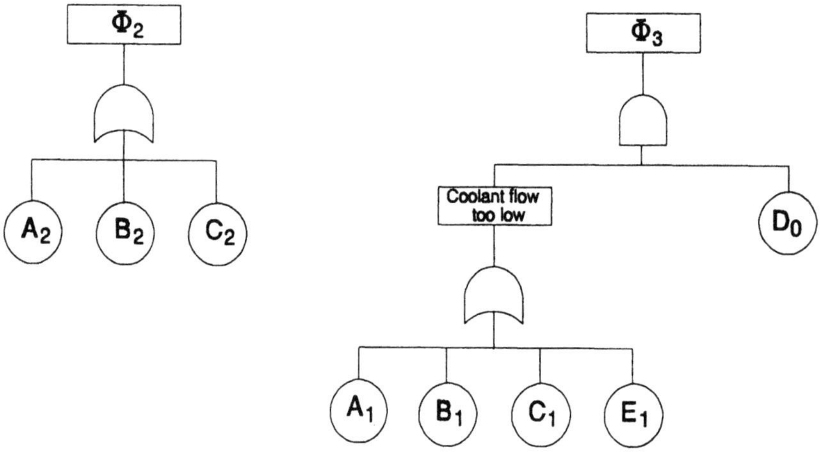

Fig. 9.22 Fault trees for Problem 9.4.

successful shutdown action (Φ_3). The fault trees for these system states are shown in Fig. 9.22 from which the expression for Φ_2 and Φ_3 can be found as:

$$\Phi_2 = A_2 + B_2 + C_2$$
$$\Phi_3 = D_0 \cdot (A_1 + B_1 + C_1 + E_1) \tag{9.27}$$

Comparison of the expression for Φ_1 and Φ_2 shows that the events are not completely mutually exclusive but have a small overlap. This is a result of the fault-tree construction process where compensation of the effect of one rare event by that of another rare event usually is ignored. However, as long as the overlapping events are all rare the effect on probability and frequency calculations is much smaller than the uncertainty in probabilistic data of the basic events and, therefore, can be ignored.

2. The fourth system state Φ_0 (successful operation) can be found as the event complementary to the union of the events Φ_1, Φ_2 and Φ_3

$$\left.\begin{aligned}\Phi_0 &= \overline{\Phi_1 + \Phi_2 + \Phi_3}\\ &= \overline{A_2 + B_2 + C_2 + (D_0 + D_1)(A_1 + B_1 + C_1 + E_1)}\\ &= \overline{(A_1 + A_2) + (B_1 + B_2) + (C_1 + C_2) + E_1}\\ &= A_0B_0C_0E_0.\end{aligned}\right\} \tag{9.28}$$

Although transitions from one system failure mode to another are possible their contributions to the frequencies of these failure modes are very small due to the fact that the failure modes themselves are rare events. In this light the occurrence frequencies of the three failure modes can be approximated as

$$f(\phi_{01}) \approx A_0B_0C_0D_1E_0[\mu(a_{01}) + \mu(b_{01}) + \mu(c_{01}) + \mu(e_{01})] \tag{9.29}$$

$$f(\phi_{02}) \approx A_0B_0C_0E_0[\mu(a_{02}) + \mu(b_{02}) + \mu(c_{02})] \tag{9.30}$$

$$f(\phi_{03}) \approx A_0B_0C_0D_0E_0[\mu(a_{01}) + \mu(b_{01}) + \mu(c_{01}) + \mu(e_{01})] \tag{9.31}$$

REFERENCES

Bossche, A. (1984) *Microelectron. Reliab.*, **24**(4), 707–15.
Bossche, A. (1986) *Microelectron. Reliab.*, **26**(3), 481–2.
Ding-Hua Shi (1981) *IEEE Trans. Reliability*, **R-30**, pp. 444–7.
Fussell, J.B. and Vesely, W.E. (1972) *Trans. Amer. Nucl. Soc.*, **15**(1).
Gadani, J.P. and Misra, K.B. (1981) *IEEE Trans. Relibility*, **R-30**, 357–61.
Huang, X.Z. (1982) *Microelectron. Reliab.*, **22**(5), 945–7.
Nahman, J.M. (1981) *IEEE Trans. Relibility*, **R-30**, 353–6.
Schneeweiss, W.G. (1981) *IEEE Trans. Relibility*, **R-30**, 362–3.
Singh, C. and Billinton, R. (1975) *IEEE Trans. Relibility*, **R-24**, 175–85.
Singh, C. (1975) *Microelectron. Reliab.*, **14**, 293–4.

CHAPTER 10

Networks and directional graphs

With the growing need for communication and distribution networks it is important to have algorithms available that allow one to calculate and assess the reliability performance of a network. The large diversity of network applications with differences in edges (bi-directional *v.* unidirectional), transported items (information *v.* product transport), load situations and routing algorithms makes it impossible to develop a single comprehensive algorithm producing exact reliability figures for each network. The problem becomes even more complex when one and the same network is used for transport of items that are different in nature. This is, for example, the case in ISDN (Integrated Services Digital Network) communication networks which will be used to transport real-time signals for which delays are unacceptable (e.g. telephone calls and TV signals) and signals for which a certain degree of irregular delays between information packages is acceptable.

Therefore, analytic network reliability algorithms always use simplified descriptions of actual networks and usually distinguish no more parameters than network graphs and edge probabilities (and in some cases also vertex probabilities). Even then exact calculations might be impossible for larger networks since the amount of computing time and/or disk space grows exponentially with network complexity. This means that the output data of these algorithms must be judged on their validity before use. With help of more dedicated algorithms that make use of Monte Carlo simulation or successive approximation methods more realistic reliability evaluations are possible at the expense of even more processing time and temporary storage capacity and reduced applicability to other networks.

Nevertheless, this chapter will be confined to analytic algorithms only. Furthermore users and owners of communication networks usually have different views on reliability; where users are mainly interested in the probability of a successful communication between a particular pair of vertices (two-terminal reliability) the network owners will be interested in the probability that successful communications between all pairs of vertices are possible (all-terminal reliability) as well.

10.1 PATHS AND CUTS

Each subset of edges that, when functional, connect a source vertex with a destination vertex constitutes a path between these vertices. Furthermore, when all edges of this path are relevant, i.e. no edge can be removed without disconnecting the path; the path is called a minimal path. In the bi-directional network graph of Fig. 10.1 the subsets e_3, e_1e_5, e_2e_6, $e_1e_4e_6$ and $e_2e_4e_5$ form minimal paths for a connection between vertices 1 and 4. The subsets $e_1e_2e_5$ and $e_1e_4e_5e_6$ are examples of non-minimal paths.

Minimal paths for a two terminal communication between vertices i and j can be generated very simply by building a tree of paths starting at vertex i. Figure 10.2 shows the tree of paths for the two-terminal communication between vertices 1 and 4 of the graph in Fig. 10.1. In order to leave out non-minimal paths each path in the tree should be free of loops, i.e. in each path a vertex may appear only once. For a directed graph with uni-directional edges (or a partly directed graph) the minimal paths are found in the same manner except that at each vertex only outgoing edges are taken into account.

Each subset of edges that, when removed (failed), prohibits any transfer between a pair of vertices constitutes a cut for the communication between these vertices. Again, when all edges of the cut are relevant, i.e. no edge can be restored (repaired) without losing the guaranteed disconnection for this pair of vertices, the cut is a minimal cut. In Fig. 10.1 the subsets of failed edges $\bar{e}_1\bar{e}_2\bar{e}_3$, $\bar{e}_3\bar{e}_5\bar{e}_6$, $\bar{e}_2\bar{e}_3\bar{e}_4\bar{e}_5$ and $\bar{e}_1\bar{e}_3\bar{e}_4\bar{e}_6$ are minimal cuts for communication between vertices 1 and 4. The set $\bar{e}_1\bar{e}_2\bar{e}_3\bar{e}_5$ is an example of a non-minimal cut.

For *undirected graphs* the minimal cuts can be generated very simply

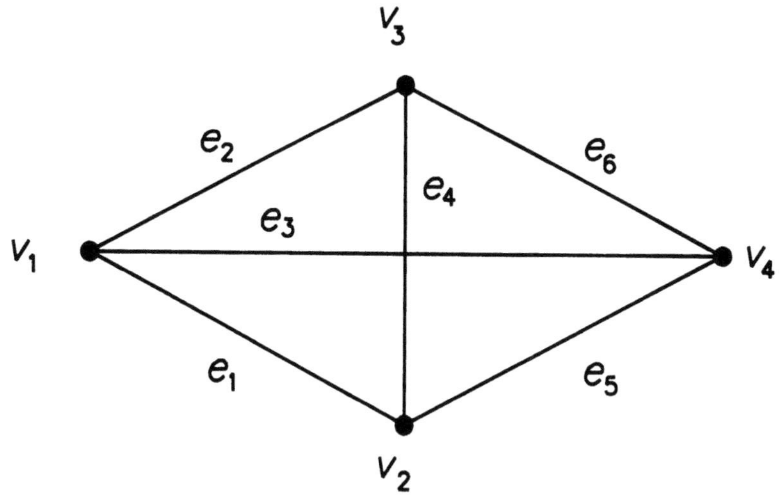

Fig. 10.1 A complete undirected (or bi-directional) graph with four nodes.

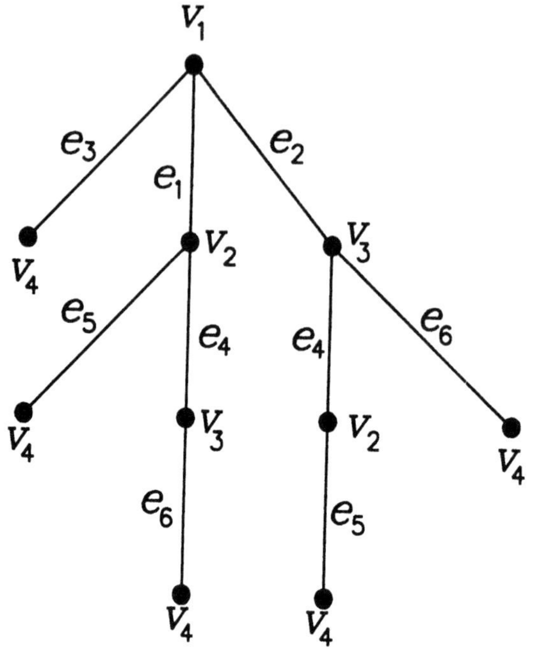

Fig. 10.2 Tree of minimal paths between vertices 1 and 4 of the graph of Fig. 10.1.

using the following procedure:

1. For all-terminal reliability take all combinations of vertex 1 with 0 or more other vertices. Each of these combinations establishes exactly one cut namely the intersection of all edges that connect one of the chosen vertices to one of the unchosen vertices (Fig. 10.3(a)). However, only those cuts that, when applied, split the original graph into exactly two subgraphs are minimal cuts. All other cuts can be ignored in reliability calculations.
2. For the two-terminal reliability between vertex i and j take all combinations of vertex i with 0 or more vertices from the set $\{N - j\}$. Each of these combinations yields a cut which is the intersection of failures of all edges from one of the chosen vertices to one of the unchosen vertices (vertex j inclusive) (Fig. 10.3(b)). Again, only those cuts that split the original graph into two separate subgraphs are minimal cuts and should be considered in calculations.

The above cut set generation procedure for two-terminal reliability also holds for *directed or partly directed graphs* when the direction is taken into account, i.e. only those edges directed from the subgraph containing the source vertex i to the subgraph containing the destination vertex j (and of course the undirected edges) are considered. The cuts found in this manner are not necessarily minimal cuts as was the case with undirected

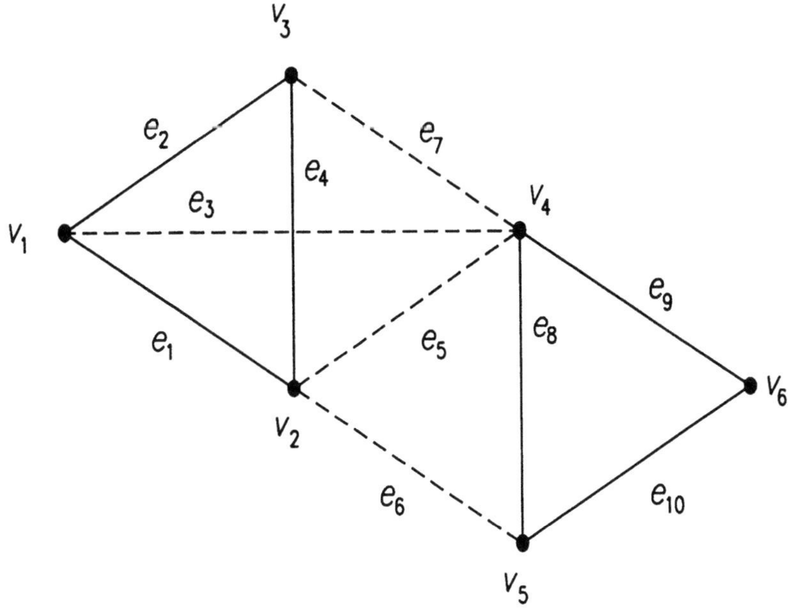

Fig. 10.3 The combination $\overline{e_3e_5e_6e_7}$ is a minimal cut for any communication between the two subgraphs since it splits the original graph into exactly two subgraphs.

graphs. Elimination of the non-minimal cuts will reduce the cut set and, hence, facilitates reliability calculations.

10.2 NETWORK RELIABILITY EVALUATION FROM PATH AND CUT SETS

As soon as a complete path set or cut set of a network is available the corresponding network reliability can be calculated. The union of all paths in a path set describes the successful connection (reliability) of the corresponding network, while the union of all cuts in a cut set describes the unsuccessful connection (unreliability) of the network. In both cases a multiple logic function results:

$$F = X_1 \cup X_2 \cup \ldots \cup X_n \qquad (10.1)$$

in which the X_i correspond to either paths (in case of reliability calculation) or cuts (in case of unreliability calculations). The most common algorithm for evaluating the probability of such a multiple logic function is the inclusion-exclusion rule introduced in Chapter 8.

As an example consider the undirected graph of Fig. 10.1. The minimal cuts that prohibit any connections between vertices 1 and 4 are: $\bar{e}_1\bar{e}_2\bar{e}_3$, $\bar{e}_3\bar{e}_5\bar{e}_6$, $\bar{e}_2\bar{e}_3\bar{e}_4\bar{e}_5$ and $\bar{e}_1\bar{e}_3\bar{e}_4\bar{e}_6$. Then the unreliability of this connection can be found by the inclusion-exclusion rule as

$$
\begin{aligned}
Pr\{F\} = {} & +\bar{e}_1\bar{e}_2\bar{e}_3 + \bar{e}_4\bar{e}_5\bar{e}_6 + \bar{e}_2\bar{e}_3\bar{e}_4\bar{e}_5 + \bar{e}_1\bar{e}_3\bar{e}_4\bar{e}_6 - \{2\bar{e}_1\bar{e}_2\bar{e}_3\bar{e}_4\bar{e}_5\bar{e}_6 \\
& + \bar{e}_1\bar{e}_2\bar{e}_3\bar{e}_4\bar{e}_5 + \bar{e}_1\bar{e}_2\bar{e}_3\bar{e}_4\bar{e}_6 + \bar{e}_2\bar{e}_3\bar{e}_4\bar{e}_5\bar{e}_6 + \bar{e}_1\bar{e}_3\bar{e}_4\bar{e}_5\bar{e}_6\} \\
& + 4\bar{e}_1\bar{e}_2\bar{e}_3\bar{e}_4\bar{e}_5\bar{e}_6 - \bar{e}_1\bar{e}_2\bar{e}_3\bar{e}_4\bar{e}_5\bar{e}_6. \qquad (10.2)
\end{aligned}
$$

This method always yields the exact answer after the calculation of the nth order intersections. However, it is a very tedious and time-consuming calculation especially when a large number of paths (cuts) is to be considered. In some cases it might even be impossible to evaluate the exact solution due to limitations on computing power. In fault-tree analysis (Chapter 9) with all low-probability (rare) events the inclusion-exclusion rule converges very rapidly to the exact answer and a first- or second-order approximation usually suffices. However, in communication networks, where a low load is economically unattractive, neither the edge availabilities nor the edge unavailabilities are so small. In this case the inclusion-exclusion calculation might even diverge initially before converging to the exact answer after the nth order terms.

Another method that will lead to an exact answer much faster is the disjoint-term method from Chapter 8 which makes all cuts in the cut-set mutually exclusive so that the cut probabilities can simply be added to find the exact answer.

Consider the complete subgraph of Fig. 10.2 again. The cuts for the all-terminal reliability case and the corresponding disjoint cuts found with the above method are given in Table 10.1.

Table 10.1 Cuts and corresponding disjoint cuts for the all-terminal reliability case

Original cuts	*Disjunct cuts*
$\bar{E}_1\bar{E}_2\bar{E}_3$	$\bar{E}_1\bar{E}_2\bar{E}_3$
$\bar{E}_1\bar{E}_4\bar{E}_5$	$\bar{E}_1\bar{E}_4\bar{E}_5\ E_2$
	$\bar{E}_1\bar{E}_4\bar{E}_5\ \bar{E}_2E_3$
$\bar{E}_3\bar{E}_5\bar{E}_6$	$\bar{E}_3\bar{E}_5\bar{E}_6\ E_1$
	$\bar{E}_3\bar{E}_5\bar{E}_6\ \bar{E}_1E_2\ E_4$
$\bar{E}_2\bar{E}_4\bar{E}_6$	$\bar{E}_2\bar{E}_4\bar{E}_6\ E_1\ E_3$
	$\bar{E}_2\bar{E}_4\bar{E}_6\ E_1\ \bar{E}_3E_5$
	$\bar{E}_2\bar{E}_4\bar{E}_6\ \bar{E}_1E_3\ E_5$
$\bar{E}_2\bar{E}_3\bar{E}_4\bar{E}_5$	$\bar{E}_2\bar{E}_3\bar{E}_4\bar{E}_5\ E_1\ E_6$
$\bar{E}_1\bar{E}_3\bar{E}_4\bar{E}_6$	$\bar{E}_1\bar{E}_3\bar{E}_4\bar{E}_6\ E_2\ E_5$
$\bar{E}_1\bar{E}_2\bar{E}_5\bar{E}_6$	$\bar{E}_1\bar{E}_2\bar{E}_5\bar{E}_6\ E_3\ E_4$

A total of 24 cut-pair comparisons were necessary to obtain the disjoint cuts. Now only 11 terms have to be calculated and added to find the exact solution, while the inclusion-exclusion rule requires 127 terms to be calculated before the exact answer is achieved.

10.3 NETWORK DECOMPOSITION

When considering a single edge of a network graph it can be either good or failed. When this edge is known to be good the graph can be simplified by contracting the two vertices of this edge. On the other hand when the edge is known to be failed it can be deleted from the graph. This offers the possibility of splitting the graph into two subgraphs, making separate calculations for each, and then combining them using Bayes' theorem as

$$R(G_{x,y}) = Pr(G_{x,y}) = e_i Pr(G_{x,y}|E_i) + \bar{e}_i Pr(G_{x,y}|\bar{E}_i) \qquad (10.3)$$

or

$$Q(G_{x,y}) = Pr(\bar{G}_{x,y}) = e_i Pr(\bar{G}_{x,y}|E_i) + \bar{e}_i Pr(\bar{G}_{x,y}|\bar{E}_i). \qquad (10.4)$$

Then these subgraphs can be decomposed again into smaller subgraphs until only single edges are left. It will be clear that a straightforward decomposition of a complex network still requires a lot of terms to be calculated, i.e. for a network of n edges a maximum of 2^{n-1} terms can be found. However, the strength of this method lies in combination with series and parallel reduction. Before each decomposition step any series combination of edges and any parallel combinations in the (sub)graph can be contracted into single edges, thus reducing the complexity of the graphs before the decomposition step. This method especially applies for the two-terminal reliability problems since then the advantage from series and parallel reductions is optimal.

EXAMPLE 10.1

Six computers are able to exchange information via the undirected network of Fig. 10.4(a). For evaluation of the probability of a successful two-terminal communication between the computers at vertices 1 and 5 the

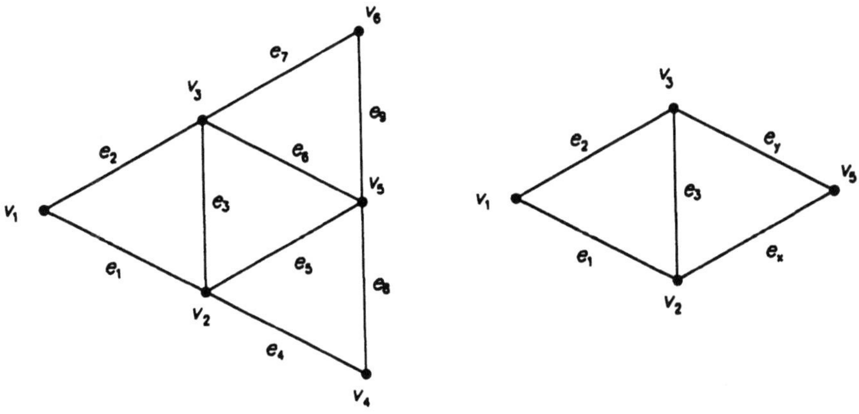

Fig. 10.4 Graph reduction by series and parallel combination of edges prior to decomposition.

graph can be simplified by series and parallel combination of edges yielding the graph of Fig. 10.4(b). Here the new edges e_x and e_y represent the subgraphs $e_5//(e_4e_8)$ and $e_6//(e_7e_9)$, respectively. No further series or parallel reduction is possible and a decomposition step using edge e_3 is necessary to evaluate the two-terminal reliability as

$$R(G_{1,5}) = e_3(e_2e_y + e_1e_x - e_1e_2e_xe_y) + \bar{e}_3(e_1 + e_2 - e_1e_2)(e_x + e_y - e_xe_y) \tag{10.5}$$

with

$$e_x = e_4e_8 + e_5 - e_4e_5e_8$$
$$e_y = e_7e_9 + e_6 - e_6e_7e_9.$$

10.4 THE BUZACOTT ALGORITHM FOR UNDIRECTED NETWORKS

The Buzacott algorithm (Buzacott, 1980) greatly depends on the cut set generation method introduced before. In this algorithm each minimal cut is combined with the event that its corresponding subgraph (the one that includes vertex 1) is connected. The resulting terms are unique and mutually exclusive and, therefore, the all-terminal reliability of the graph can be calculated simply by adding the probabilities of all terms.

The disconnection (unreliability) of a graph is calculated as

$$Q(G) = C_{0,N} + \sum_{r=1}^{n-1}\sum_{i} Pr[G_r^i]Pr[C_{r,G}^i], \tag{10.6}$$

where G is the total graph with n vertices, G_r^i is the ith subgraph consisting of vertex 1 and r other vertices, $C_{0,N}$ is the minimal cut separating vertex 1 from all other vertices, C_r^i is the minimal cut separating subgraph G_r^i from all other vertices in G.

As can be seen from formula (10.3), the availabilities of all the subgraphs G_r^i are used and should be evaluated first. Therefore, this formula is applied first to calculate the unreliabilities of all two-vertex subgraphs G_1^i which then can be used to evaluate the unreliabilities of all three-vertex subgraphs and so on. Figure 10.5 shows the subsequent terms to be calculated to evaluate the all-terminal unreliability of the four-vertex complete graph depicted in Fig. 10.1. Each node corresponds with a subgraph.

10.5 NETWORKS WITH UNRELIABLE EDGES AND VERTICES

In the previous paragraphs it was assumed that network connections failed due to edge failures only. However, in practice vertices may fail as

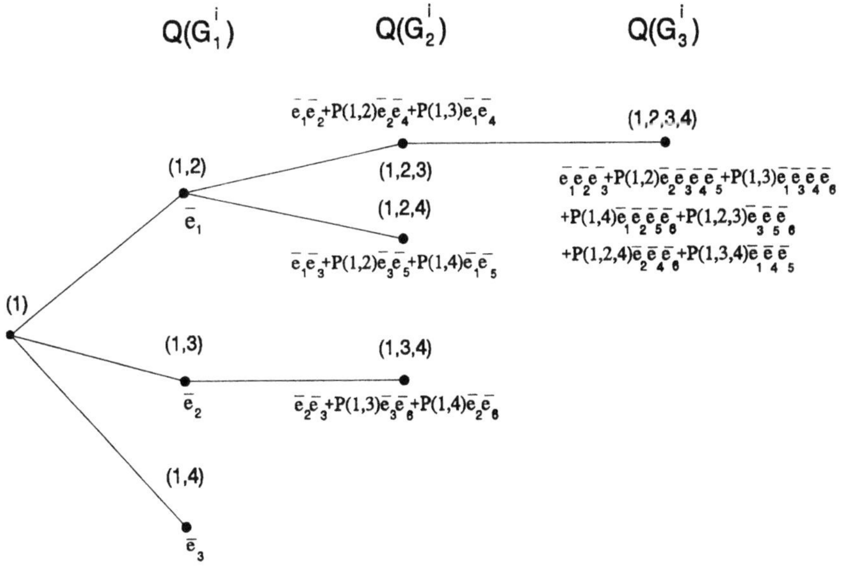

Fig. 10.5 Tree showing the subsequent terms to be calculated by the Buzacott algorithm for the evaluation of the all-terminal unreliability of the four-vertex complete graph depicted in Fig. 10.1.

well, e.g. when the computer system of an electronic telephone exchange fails. The path set approach is the best choice for this problem since the number of minimal paths is not affected; only the number of items in each path increases. For example for the two-terminal reliability between vertices 1 and 4 of Fig. 10.1 the path set now becomes: $v_1e_3v_4$, $v_1e_1v_2e_5v_4$, $v_1e_2v_3e_6v_4$, $v_1e_1v_2e_4v_3e_6v_4$ and $v_1e_2v_3e_4v_2e_5v_4$.

10.6 WORKED PROBLEMS

PROBLEM 10.1

Consider the network depicted in Fig. 10.6.

1. Reduce the network by series-parallel reduction as much as possible for the two-terminal communication between nodes v_1 and v_5.
2. Find the minimal cut set and evaluate the two-terminal reliability for this pair of nodes when all basic edges have a failure probability of 0.01.
3. Evaluate the two-terminal reliability for node pair 1–5 again when the nodes may also fail with probability 0.01.

SOLUTION

1. The edges e_4, e_5 and e_7 can be reduced to a single edge e_5' with unreliability

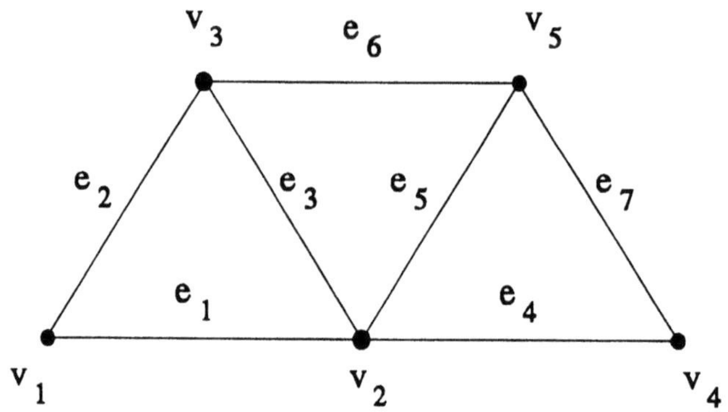

Fig. 10.6 Network of Problem 10.1.

$$\bar{e}_5' = \bar{e}_5(1 - e_4e_7).$$

2. The minimal cuts for the reduced network are

$$\bar{e}_1\bar{e}_2,\ \bar{e}_5'\bar{e}_6,\ \bar{e}_1\bar{e}_3\bar{e}_6 \text{ and } \bar{e}_2\bar{e}_3\bar{e}_5'.$$

Then in a first-order approximation the two-terminal reliability can be evaluated as

$$\begin{aligned} R(G_{1,5}) &= 1 - Q(G_{1,5}) \\ &\approx 1 - (\bar{e}_1\bar{e}_2 + \bar{e}_5'\bar{e}_6 + \bar{e}_1\bar{e}_3\bar{e}_6 + \bar{e}_2\bar{e}_3\bar{e}_5') = 0.9998. \end{aligned}$$

3. Now the edges e_4, e_5 and e_7 together with vertex e_4 can be reduced to edge e_5'' with reliability $e_5'' = 1 - \bar{e}_5(1 - e_4v_4e_7)$.

Then the minimal paths for the reduced network are $v_1e_1v_2e_5''v_5$, $v_1e_2v_3e_6v_5$, $v_1e_1v_2e_3v_3e_6v_5$ and $v_1e_2v_3e_3v_2e_5''v_5$.

A first-order approximation with the inclusion-exclusion rule is inadequate because the probabilities (reliabilities) are too high for a fast convergence, Therefore, the calculation has to be completed upto the fourth-order terms. This results in:

$$\begin{aligned} R(G_{1,5}) = &+v_1e_1v_2e_5''v_5 + v_1e_2v_3e_6v_5 + v_1e_1v_2e_3v_3e_6v_5 \\ &+ v_1e_2v_3e_3v_2e_5''v_5 - v_1e_1v_2e_2v_3v_5e_5''e_6 \\ &- v_1e_1v_2v_3e_3v_5e_5''e_6 - v_1e_1v_2e_2v_3e_3v_5e_5'' \\ &- v_1e_1v_2e_2v_3e_3v_5e_6 - v_1v_2e_2v_3e_3v_5e_5''e_6 \\ &- v_1e_1v_2e_2v_3e_3v_5e_5''e_6 + 4v_1e_1v_2e_2v_3e_3v_5e_5''e_6 \\ &- v_1e_1v_2e_2v_3e_3v_5e_5''e_6 \\ = &\ 0.979. \end{aligned} \tag{10.7}$$

PROBLEM 10.2

Consider the (partial) telephone network of Fig. 10.7. This simple network consists of three exchanges (big nodes) each having three subscribers

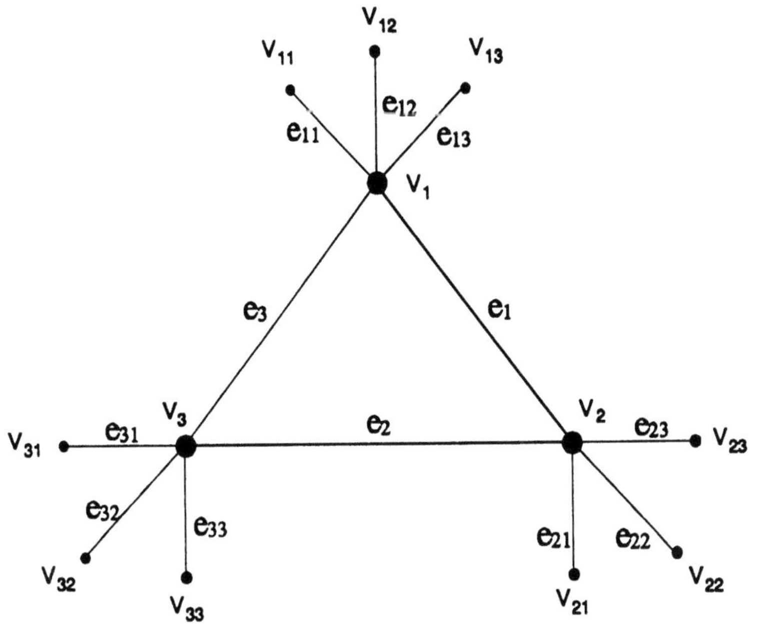

Fig. 10.7 Network of Problem 10.2.

(small nodes) connected. All subscriber edges have a failure probability of 10^{-4} while the interlocal edges between the exchanges have an unavailability of 10^{-2} (since these lines are used for other types of communication as well).

Evaluate the all-terminal reliability of this network.

SOLUTION

Since all subscribers are connected directly to their local telephone exchange the all-terminal reliability problem can be split into two independent probabilities: i.e. all subscriber edges must be available *and* the three exchanges must be connected. Then the all-terminal network reliability can be evaluated as

$$R(G) = Pr(\text{all subscriber lines available})Pr(\text{all exchanges connected})$$

$$= \left[\prod_{i=1}^{3}\prod_{j=1}^{3} e_{ij}\right][e_1e_2 + e_1\bar{e}_2e_3 + \bar{e}_1e_2e_3]$$

$$= 0.999\,10 \times 0.999\,70 = 0.998\,80. \tag{10.8}$$

REMARK

Although the non-redundant subscriber lines have a higher individual reliability their combined contribution to network unavailability is approximately three times larger than that of the redundant interlocal edges.

PROBLEM 10.3

Use series-parallel reduction and network decomposition techniques to evaluate the two-terminal reliability $R(G_{1,9})$ of the network shown in Fig. 10.8 when all edges have a unreliability of 10^{-2}.

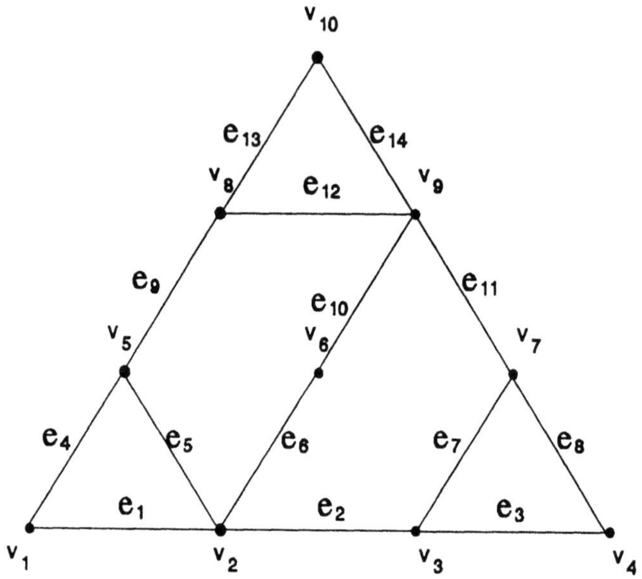

Fig. 10.8 Network of Problem 10.3.

SOLUTION

Edges in series can be reduced to a single edge with an reliability that equals the product of the reliabilities of the respective independent edges, while for reduction of parallel edges to a single edge the unreliabilities

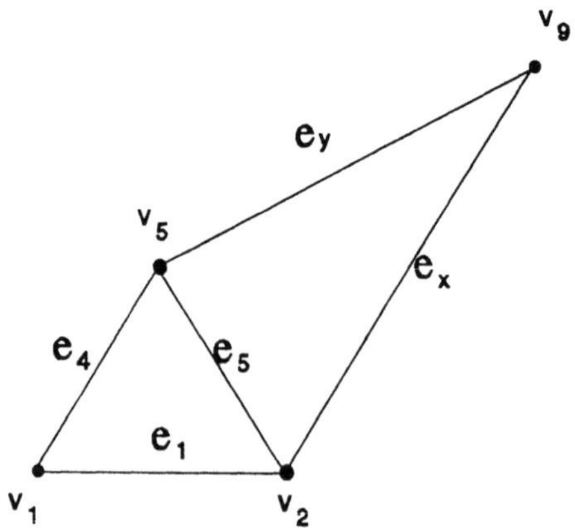

Fig. 10.9 Reduced network of Problem 10.3.

can be multiplied to obtain the unreliability of the resulting edge. With these rules the network between the vertices v_1 and v_9 can be reduced to the network of Fig. 10.9 where the unreliabilities of the edges e_x and e_y can be evaluated as

$$\bar{e}_x = [1 - e_5 e_{10}][1 - e_2 e_{11}(1 - \bar{e}_7(1 - e_3 e_8)] \approx 4 \times 10^{-4} \quad (10.9)$$

and

$$\bar{e}_y = 1 - e_9[1 - \bar{e}_{12}(1 - e_{13} e_{14})] \approx 1.02 \times 10^{-2}. \quad (10.10)$$

Then, using edge e_5 in Fig. 10.9 as the key edge for decomposition, the two-terminal reliability $R(G_{1,9})$ can be found as

$$R(G_{1,9}) = e_5[1 - \bar{e}_1 \bar{e}_4][1 - \bar{e}_x \bar{e}_y] + \bar{e}_5[1 - (1 - e_4 e_y)(1 - e_1 e_x)] \approx 0.999\,895. \quad (10.11)$$

REFERENCES

Buzacott, J.A. (1980) *Networks*, **10**, 311–27.

Klaassen, K.B. and van Peppen, J.C.L. (1988) *System Reliability, Concepts and Applications*, Edward Arnold, London.

Satyanarayana, A. and Prabhakar, A. (1978) *IEEE Trans. on Reliability*, **R-27**(2), 82–100.

Shooman, M.L. (1968) *Probabilistic Reliability: An Engineering Approach*, McGraw-Hill, New York.

CHAPTER 11

Maintained System Reliability

The definitive work on this subject is that of Ascher and Feingold in their many papers culminating in a book entitled *Repairable Systems Reliability: Modelling, Inference, Misconceptions and their Causes* (Ascher and Feingold, 1984). It is not our intention to steal their thunder nor to present such a detailed and mathematically rigorous text as theirs. However, we set out to write a book covering all aspects of system reliability, availability and productiveness. The special considerations needed for maintained systems are extremely important and generally ill-understood even by some quite experienced practitioners.

11.1 PROCESS AND DISTRIBUTION DISTINGUISHED

In Chapter 1 we discussed the two forms of bathtub curve, one for components and the other for repairable, maintained systems, also their different origins and the confusion between them and between the concepts of hazard rate and ROCOF both loosely referred to as 'failure rate'. Readers may wish to review that section before studying the rest of this chapter.

The system bathtub curve traces the varying instantaneous mean ROCOF of a population of nominally identical systems. Each individual system will behave differently. The time to failure of each component is essentially a random variable with a distribution which can be estimated from data. The ROCOF for an individual system is therefore a wavy variant of the smooth curve obtained by averaging them all. The ROCOF at any instant T in the life of the system (measured from new or entry into serious service), $\lambda(T)$, is the sum of the instantaneous hazard rate functions $z_i(t_i)$, where each t_i is measured from the last renewal of that component. For a system of n components

$$\lambda(T) = \sum_{i=1}^{n} z_i(t_i) \tag{11.1}$$

The components have failure p.d.f.s, $f_i(t_i) i = 1 \ldots n$ corresponding to the $z_i(t_i)$, and related to their individual lives t_i, whereas the system as a whole suffers a process, which may be modelled as a NHPP (non-homogeneous Poisson process) related to its age from new. All renewals of components, including preventive renewals, affect the value of $\lambda(T)$, which by convention is taken to be the average instantaneous value for a population of nominally identical systems.

11.1.1 Constant ROCOF

If the $f_i(t_i)$ are unchanging with T, as is usually taken to be the case in the mid-portion of the system bathtub curve, then the individual distribution have constant means defined by

$$\theta_i = \int_0^\infty t_i f_i(t_i)\mathrm{d}t_i. \tag{11.2}$$

In this case only, the NHPP becomes a homogeneous Poisson process or HPP with constant rate λ where in the absence of PM

$$\lambda = \sum_{i=1}^{n}(1/\theta_i). \tag{11.3}$$

That is to say, the probability of exactly r failures in an interval $\Delta T = T_2 - T_1$ is given by

$$P(r,\Delta T) = [(\lambda\Delta T)^r/r!]\exp(-\lambda\Delta T). \tag{11.4}$$

When $r = 0$

$$P(0,\Delta T) = \exp(-\lambda\Delta T). \tag{11.5}$$

$P(0,\Delta T)$ the probability of zero failures is the *conditional* interval reliability of the system over *any* period of length ΔT *given* that it was in working order at the start of the interval, T_1. For an HPP, $P(0,\Delta T)$ is always the same whatever the value of T_1, *by definition of an HPP*. For an NHPP with a constant portion, it may be true between, say, T_A and T_B, the limits of that flat portion. For an NHPP where λ is a function, $\lambda(T)$, of T, the conditional interval reliability is

$$P(0,T_1,T_2) = \exp\left[-\int_{T_1}^{T_2}\lambda(T)\mathrm{d}T\right]. \tag{11.6}$$

The integral in equation (11.6) is the expected number of failures in the interval. Equations (11.5) and (11.6) are reminiscent of the reliability function of the exponential distribution. However, the two functions have different meanings as well as different time-variables. One is conditional, the other a simple probability; one refers to the whole system, the other to a component.

It is true that such an HPP does imply exponentially distributed intervals between successive failures. Thus the s.d. of such intervals is equal to their mean. For computational purposes and *provided the maintenance schedule and other conditions of operation are not changed*, the mathematics of the exponential distribution can be used. As we explained in the chapter on the bathtub curve, preventive maintenance (PM) may be taken to simply lower the value of λ by eliminating a proportion of the failures. Simulation suggests it will also make Drenick's waves more regular in both frequency and amplitude, but apart from special cases it will not be worthwhile to model $\lambda(T)$ as a waveform. In a truly random failure regime, by contrast, it is not possible to alter the constant hazard rate z by taking precautionary measures such as preventive maintenance. In such a regime all the failures result from accidents, coincidences of relatively rare events and what insurers call 'Acts of God'. The point is that there is nothing the operator can do to prevent them within the rules. Strictly, the term Poisson process, whether homogeneous or not, implies true randomness of events and helplessness of the operator to affect the outcome. We should therefore really refer to what occurs to repairable and maintained systems as pseudo-HPP or pseudo-NHPP.

It is important for many reasons that the previous woolly thinking about the failure mathematics of repairable and maintained systems be clarified, but perhaps the most important is the disastrous economic effect it can have upon maintenance practice. Almost every book on maintenance optimization concludes mathematically that if the failure rate is constant, preventive maintenance should be abandoned. As we have seen though, constant $\lambda(T)$ does not imply that all the $z_i(t_i)$s are also constant! Preventive schedule optimization is a subject for another book. We will confine our remarks here to three observations:

1. The term 'failure rate' should be banned from serious reliability literature. If forced to use either 'hazard rate function', $z(t)$ or 'rate of occurrence of failures', ROCOF, $\lambda(T)$ analysts would remain conscious of their essentially different nature.
2. Very often the ROCOF is constant because, and only because, there is repair and maintenance. The intensity of PM and the scope of repairs, (particularly as to what is regarded as repairable), determine the perceived ROCOF and limit the period of its constancy.
3. The PM schedule and the scope for repairs need to be determined or estimated on *economic* grounds in most cases. Anything can be made to last for any time we choose by varying these factors, but there is only one economically optimal policy, which generally we should strive to find, whilst acknowledging that in practice we never can get it absolutely right before the system becomes obsolete. This is partly because we are trying to hit a moving target.

11.1.2 Effect of maintenance

The effect of maintenance upon λ or $\lambda(T)$ depends upon the combined effects on the θ_i. If, to take a common example, we have a series system in which some components are subject to age renewal (renewal at age t_i^* from last renewal or failure and at failure), and λ is constant, then the ith component's contribution to λ will be reduced from $1/\theta_i$ to

$$1/\theta_i^* = \lambda_i^* = F_i(t_i^*)/[t_i^* R_i(t_i^*) + \int_0^{t_i^*} t_i f_i(t_i)\mathrm{d}t_i]$$

$$= Pr\ \text{(failure cycle)/(mean cycle time, renewal to renewal)} \quad (11.7)$$

and the combined ROCOF despite PM would be

$$\lambda^* = \sum_{i=1}^{n}\left(F_i(t_i^*)/[t_i^* R_i(t_i^*) + \int_0^{t_i^*} t_i f_i(t_i)\mathrm{d}t_i]\right) = \sum_{i=1}^{n}\lambda_i^*. \quad (11.8)$$

In equation (11.7), θ_i^* represents the conditional mean cycle time between renewals (whether for failure or on age) divided by the probability that a cycle ends in failure. The denominator will be familiar to those who have studied PM schedule optimization (see for example Jardine, 1981). Age renewal is usually optimized for cost. If the mean costs of a failure and of PM are, respectively, C_{Fi}, C_{Mi} where $C_{Fi} > C_{Mi}$ then the mean cycle cost for item i is

$$C_{Fi}F_i(t_i^*) + C_{Mi}R_i(t_i^*) \quad (11.9)$$

and the cost rate (mean long-term rate of expenditure per unit time) to be minimized is

$$c_i(t_i^*) = \operatorname*{Min}_{t^*}[C_{Fi}F_i(t_i^*) + C_{Mi}R_i(t_i^*)] \Big/ \left[\int_0^{t_i^*} t_i f_i(t_i)\mathrm{d}t_i + t_i^* R_i(t_i^*)\right]. \quad (11.10)$$

If the $t_i^*, i = 1, \ldots, n$ are all optimal, the resulting λ^* represents a practical economic lower bound on the system overall ROCOF. In practice greater financial savings can be made by grouping PM schedules by the calendar rather than individual ages and taking opportunities offered by stoppages caused by failures to advance PM, which are known as *block* and *opportunity* maintenance, respectively. It is useful though, to find this bound as a guide to where improvements in a planned or existing system can be economically justified to meet a production target (Chapter 6) or an availability specification.

11.2 RELIABILITY GROWTH – TESTING TO MEET TARGETS

In the previous section and in Chapter 1 (bathtub curve) we theorized that mean ROCOF, $\lambda(T)$ in a complex system would settle to a constant value if all the conditions of operation, repair and PM were kept constant and would remain constant until long-term wear out modes of failure started to occur. This is the constant portion of the bathtub curve.

Drenick's waves (Fig. 1.4) are associated with individual manifestations of the system. Their pattern is random, unique and unpredictable, whilst their constant mean, λ, can in theory be predicted from perfect data in respect of all the components $i = 1, \ldots, n$.

There are two common circumstances in which $\lambda(T)$ is not expected to be constant. One is during testing and development of new systems, which is discussed in this section, and the other is increasing ROCOF and failure costs in systems where sufficient maintenance to hold λ constant is not operationally feasible, and much PM is left to designated overhaul periods. Overhaul is discussed in the next section.

System reliability development, like performance development, often requires a program of testing and modification followed by more testing in order to meet a pre-conceived specification. Such programs are known as test analyse and fix or TAAF programs and are much employed by the military. This is not surprising for three reasons which also apply to civilian equipment in some cases.

1. Military equipment is often at the cutting edge of technological innovation.
2. The customer hopes never to have to use the equipment, but if it is used it must work. Also, its ability to work must be credible to deter potential enemies.
3. Military equipment must operate under the most rugged and varied conditions and so needs are reserve of reliability and resistance to climatic and other extremes.

The war following Iraq's invasion of Kuwait in which the allies moved against Saddam Hussein's forces to protect their oil supplies, was a triumph for reliability engineering and in particular for the TAAF principle. The very sophisticated technology worked sufficiently well for the ground war to be won decisively in only four days with a disparity in military casualties of three orders of magnitude and with low civilian loss of life compared to earlier wars. Although evidence emerging since the war has tended to play down the scenario of pinpoint accuracy and high reliability presented to reporters at the time of the war, nevertheless such sophisticated technology would probably not have been usefully reliable at all had its reliability not been extensively developed by TAAF.

In the mid 1960s J.T. Duane (1964) analysed the growth in reliability of several systems under a regime of constant redesign effort. He found that *if the resources devoted to improvement remained constant* over the TAAF program then the MTBF continued to improve, subject to diminishing returns. Plotting the cumulative MTBF at successive equal intervals one obtains an empirical power function model for the law of growth in almost every case. The cumulative MTBF is simply the total test hours to date divided by the number of failures. In TAAF, failures occur, their causes are investigated and preventive measures in the form of modifications to the design, and alterations to the quality control arrangements in manufacture, or failing those, amendments to the PM schedule, are introduced successively as testing continues. It is generally found that the cumulative mean time between failures (CMTBF) follows a power law thus

$$\text{CMTBF} = kT^{\alpha} \tag{11.11}$$

Now

$$\text{CMTBF} \equiv T/F, \tag{11.12}$$

where F is the number of failures in T. It follows that

$$F = T^{1-\alpha}/k$$

and so

$$\lambda(T) \equiv \mathrm{d}F/\mathrm{d}T = (1 - \alpha)/kT^{\alpha}. \tag{11.13}$$

$\mathrm{d}F/\mathrm{d}T$ is the *current* ROCOF, $\lambda(T)$ so the model for current MTBF, which is $1/\lambda(T)$ by definition is

$$\theta(T) = kT^{\alpha}/(1 - \alpha). \tag{11.14}$$

Thus the instantaneous or current MTBF is the CMTBF multiplied by $1/(1 - \alpha)$. Crow reparameterized (11.13) above so that the expression for $\lambda(T)$ resembled a Weibull hazard rate function, i.e.

$$\lambda(T) = \beta T^{\beta-1}/\eta^{\beta}$$

or

$$\lambda(T) = \delta\beta T^{\beta-1} \quad \text{(Crow)}. \tag{11.15}$$

In our opinion as well as that of Ascher and Feingold (1984), Crow has added to the confusion surrounding NHPPs and distributions by calling this a 'Weibull process'. Crow himself is definitely not confused, but we think his terminology confuses others. Ascher and Feingold point out however that acceptance of Duane's original *entirely empirical* postulate leads to an expression for the expected number of failures in 0 to T as a simple power function of T, i.e. integrating (11.15).

$$\bar{F}(T) = (T/\eta)^{\beta}. \tag{11.16}$$

This resembles the cumulative hazard function $H(t)$ for a Weibull distribution, but is actually the mean of a NHPP in which the probability of j failures in 0 to T is

$$P_2\{F(T) = j\} = (T/\eta)^{\beta}\mathrm{e}^{-(T/\eta)^{\beta}}/j!. \tag{11.17}$$

Note that $\bar{F}(T)$ is *not* the renewal function of a Weibull distribution. Nor is η a 'characteristic life' in this case but simply a parameter of the NHPP expression having the dimension of time.

The Duane model is usually operated graphically as follows. Plot T/F versus T on log/log graph paper and put the best straight line through the points by eye to obtain an estimate of CMTBF. (Duane actually plotted F/T to get a plot of what we now call ROCOF, but we feel that the method is marginally easier to follow as growth of MTBF.) The gradient of this line is the estimate of α and can easily be found by linear measurement horizontally from a point on the line a distance say h and then vertically to rejoin the line at a distance say v. Then take α as v/h. The other parameter k can be found then from any two points (on the line rather than data points). The line representing the instantaneous MTBF, $\theta(T)$ is then drawn by taking any two well-spaced points on the original CMTBF line and multiplying by $1/(1 - \alpha)$ to obtain a parallel line which will be above the CMTBF line if the MTBF is increasing and below it if there is deterioration rather than improvement in the system reliability.

Crow (1974) found maximum likelihood estimators for his parameters which equate to $1/k$ and $1 - \alpha$ in the notation we have used. In a later paper (Crow, 1975) he obtained, by extensive simulation, percentage points of the distribution of the ROCOF divided by its maximum likelihood estimate at the time of the jth failure from $T = 0$, i.e. $\lambda(T)/\hat{\lambda}(T)$, which of course is equal to $\hat{\theta}(T)/\theta(T)$. Crow's estimates were confirmed mathematically by Lee and Lee (1978). Thus, it is possible to estimate not only $\theta(T)$ but also its confidence limits.

However, it is not usually done that way. Usually, the TAAF program is followed by supplementary test time after the design has been frozen. This is because engineering actions may be more or less effective than the last entries of the growth trial would suggest statistically. It is assumed that the ROCOF will remain constant for this last period of testing,

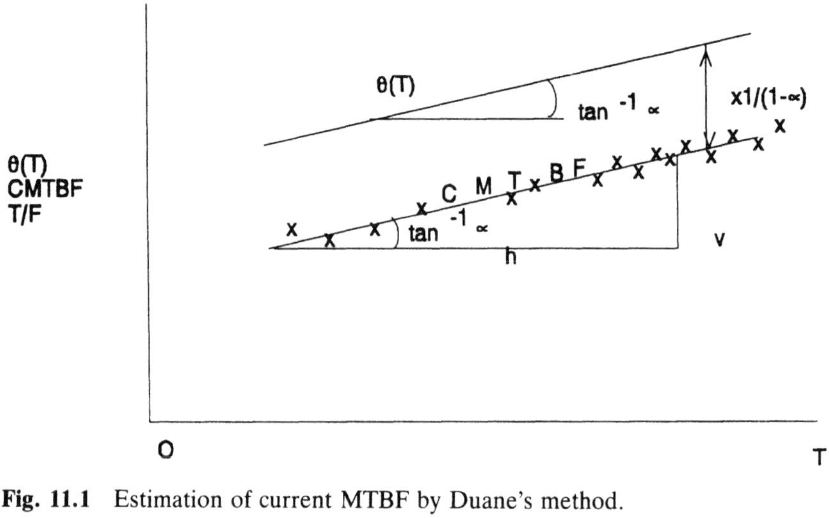

Fig. 11.1 Estimation of current MTBF by Duane's method.

usually called *acceptance testing*. An estimate and confidence limits of λ can be obtained from the acceptance test alone as in the following example.

EXAMPLE 11.1

In an acceptance test eight failures occur in a total of 500 test hours. (This may seem a lot but even acceptance tests are often conducted under harsher conditions than anticipated for actual service in order that some failures do occur.)

$$\text{Point estimate } \theta = 500/8 = 62.5\,\text{h}$$
$$\lambda = 1/\theta = 0.016.$$

Lower 90% double-sided limit of λ

$$\lambda_L = \chi^2_{0.95,\,2(8+1)}/2 \times 500 = 9.39 \times 10^{-3}/\text{h from tables.}$$

Upper limit of θ

$$\theta_u = 1/\lambda_L = 106.5\,\text{h}.$$

Upper limit of λ

$$\lambda_U = \chi^2_{0.05,\,16}/2 \times 500 = 0.0263/\text{h}.$$

Lower limit of θ

$$\theta_L = 1/\lambda_U = 38.03\,\text{h}.$$

We are thus 90% confident that λ and θ lie between these limits. The assumption underlying the above method of estimation is that the likelihood function of λ, $f(\lambda)$, has a gamma distribution with parameters F and $1/T$, i.e.

$$f(\lambda) = T(\lambda T)^{F-1}e^{-\lambda T}/(F-1)!. \tag{11.18}$$

Now the inverse distribution function of the gamma (Γ) is related to χ^2, thus

$$G_\alpha(\Gamma : b, c) = \chi^2_{1-\alpha,\,2c}/b. \tag{11.19}$$

The two extra degrees of freedom in the lower limit of λ arise because in the general case T does not coincide with a failure. If it does then in both limits the d.f. are $2F$. Note also that (11.18) is the relationship in Poisson form for the probability of F failures in T when the ROCOF is λ. They are not the same thing because in this case λ is the 'variable'. λ is not really variable, it is a constant about whose value we are doubtful on the evidence of only F failures in a limited time T.

Our purpose here is to discover whether a specification has been met, such as 95% confidence that the MTBF exceeds 40 h. In this case we have insufficient evidence to be 95% confident but enough to be 90% confident so we are nearly there.

11.2.1 Bayesian estimation method

Suppose now that we do not want to continue testing to obtain the additional hours. The Chief Engineer asks why we cannot make use of all the data we accumulated during the development testing and its Duane plot. The fact is that we can. The development testing work is the source of our growing feeling that the product is in fact within specification so we need an organized mathematical way of bringing it into the final calculation. It is usual to proceed as follows.

Suppose that during the development testing the design was modified several times to eliminate or reduce the frequency of the more pernicious failure modes. We now use the data arising before and after such design modifications to judge their engineering success. Suppose there were F failures in T_1 during development. We could just say that our best estimate of θ at the end of testing was $\theta(T)$ from the Duane plot and that therefore in a further period T_1 we would expect the number of failures to be on average

$$F_1 = F\text{CMTBF}/\theta(T_1) = (1 - \alpha)T_1^{1-\alpha}/k. \tag{11.20}$$

Suppose this is followed by our previously discussed acceptance trial in which eight failures occurred in 500 hours, i.e. $F_2 = 8$, $T_2 = 500$. The two data sets can then be combined so that $f(\lambda)$ has parameters

$$b = 1/(T_1 + T_2), \quad c = F_1 + F_2.$$

As with any Bayesian procedure, weightings of the prior and new data are subjective, that is, judgement may be required to alter the 'natural' weighting used here.

Suppose that during development testing to obtain the Duane plot, 25 failures occurred in a 1000-hour program. However the Duane plot had a definite slope, $\alpha = 0.2$. First we find k from

$$\text{CMTBF} = kT^{\alpha}$$

$$1000/25 = k \times 1000^{0.2}$$

$$k = 40/3.981 = 10.05.$$

Then find $\theta(T)$ from

$$\theta(T) = kT^{\alpha}/(1 - \alpha) = \text{CMTBF}/(1 - \alpha)$$
$$= 40/0.8 = 50 \text{ hours.}$$

The expected number of failures would therefore be $25 \times 40/50 = 20$ in 1000 hours rather than 25. This gives $b = 1/1500$ and $c = 28$.

The next step is to combine the two gamma estimates giving a posterior estimating function for λ of

$$f_p(\lambda) = 1500(1500\lambda)^{27}e^{-1500\lambda}/27!.$$

The posterior central estimate (mean) of λ is then $\lambda_p = 28/1500 = 0.018\,67$/h and of course $\theta_p = 1/\lambda_p = 53.57$ hours.

The 90% two-sided limits are then

$$\lambda_{LP} = \chi^2_{0.95,58}/2 \times 1500 = 41.503/3000 = 0.013\,83$$

$$\theta_{UP} = 1/0.013\,83 = 72.28$$

$$\lambda_{UP} = \chi^2_{0.05,56} = 74.451/3000 = 0.0248$$

$$\theta_{LP} = 1/0.0248 = 40.29\,\text{h}.$$

In this example, the specification has been met but only by a very small margin. Note that although the prior central estimate was less than that of the acceptance trial, the additional information still refined the lower limit sufficiently to meet the specification.

The putative number of failures in the prior estimate need not be based on the Duane plot alone. It is common practice to allow for modifications made after the development testing and before the final test. Provided that failure modes against which design action is taken at this point do not recur during acceptance testing, and that the modifications do not themselves introduce new failure modes or exacerbate old ones, it is fair to reduce the putative number of failures in F_1 by say nine tenths of the number of failures in the mode in question. If these conditions are not met, then development testing should not be regarded as complete until fresh modifications are proved successful. No previous work need be lost in doing this; one simply continues the Duane plot.

11.2.2 Sequential testing

The method, developed by Wald (1947), for testing batches of product with minimum average number of separate tests required to prove that a specification has been met, has been adapted for acceptance testing of designs. In principle, every hour spent without failure increases confidence that the specification has been met and every failure is a set-back. We can never be quite sure that a specified MTBF will be exceeded, nor on the other hand that we will ever reach it. The method is actually worked in terms of ROCOF, λ, rather than MTBF.

We choose a low ROCOF λ_1 which is to be accepted with high probability $(1 - \alpha)$ and a higher value λ_2 which is to be rejected with high probability $(1 - \beta)$. Frequently α and β are equal. λ_2 and $(1 - \beta)$ may be the specified conditions, i.e. it is required to show that $\lambda \leqslant \lambda_2$ with confidence $(1 - \beta)$. λ_1 and α are chosen to represent the lowest ROCOF estimate available, assessed from experience and engineering judgement, and a low probability of rejecting such a design.

The value $\hat{\lambda}$ that the designers and reliability engineers actually expect to be achieved should lie between λ_1 and λ_2. In MIL-STD 781, λ_1 is taken as the 'design' ROCOF so that λ_2/λ_1 becomes the 'design ratio,' i.e. to ensure that the specification is met using estimates of component ROCOFs perhaps from MIL-HBK 217. The situation may be represented as an operating characteristic curve of probability of accepting the design, P_A, versus ROCOF, λ.

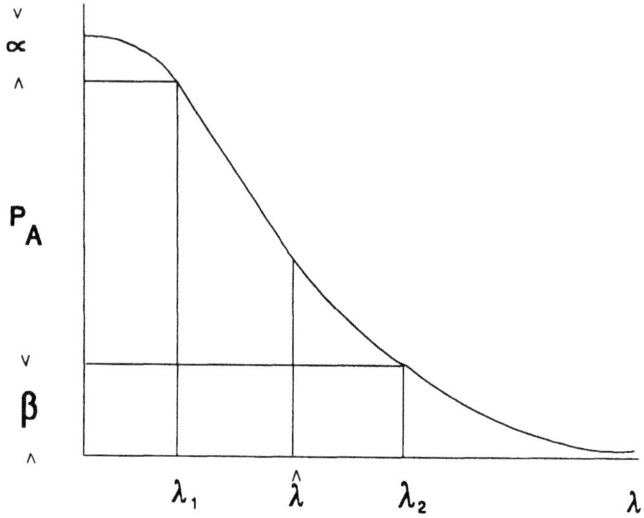

Fig. 11.2 Operating characteristic curve for design acceptance on the basis of ROCOF λ.

Then calculate

$$h_0 = \log_e[(1 - \alpha)/\beta]/(\lambda_2 - \lambda_1)$$

$$h_1 = \log_e[(1 - \beta)/\alpha]/(\lambda_2 - \lambda_1)$$

$$S = \log_e(\lambda_2/\lambda_1)/(\lambda_2 - \lambda_1). \tag{11.21}$$

Then, on a graph of number of failures, F, versus cumulative test time, T, draw two lines of slope I/S with equations

$$\text{Accept line: } t = SF + h_0$$

$$\text{Reject line: } t = SF - h_1.$$

The plot may be truncated with very little loss of confidence, at $T_{max} = Sr$ and $F_{max} = 3r$, where r is found by inspection of the χ^2 tables from

$$\lambda_1/\lambda_2 \approx \chi^2_{1-\alpha,2r}/\chi^2_{\beta,2r}, \tag{11.22}$$

i.e. r is the rejection number for the single sampling plan with the same α, β, λ_1 and λ_2. Frequently only one limit applies.

The plot starts at the origin and continues until one of the lines is

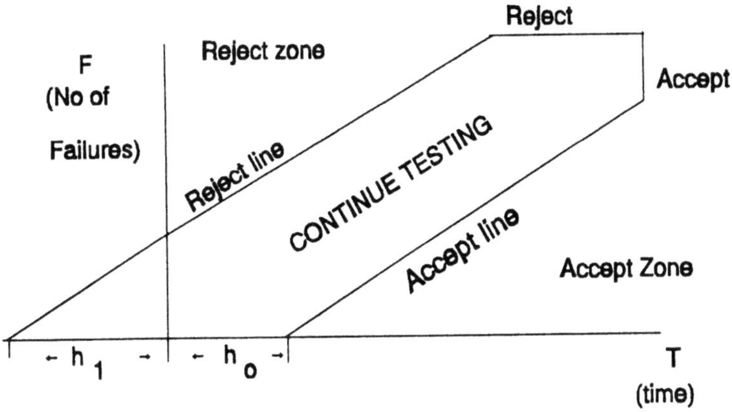

Fig. 11.3 Waid sequential test plan – basic diagram.

crossed and the design is accepted or rejected. If accepted, a check is made against the specification, i.e. check that

$$\lambda_2 \leqslant \chi^2_{\beta,2(F+1)}/2T.$$

Lloyd and Lipow (1964) have suggested that a sequential test plot be kept during development testing as an alternative to or in addition to a Duane plot and the origin moved up and rightwards successively to see whether the most recent sequence of failures makes the design acceptable. This is most easily achieved either by computer graphics or by constructing a clear plastic mask of the Wald test plan to place over the plot. As development proceeds and λ is reduced by altering the design, the gradient of the plot, which on average is a function $\lambda(T)$ of T, falls continuously and then becomes constant when all the economically corrigible failure modes have been designed out. If at this stage, the design is still in the reject zone, it should be abandoned and a fresh design approach taken. If in the continue testing zone then testing can continue as a formal acceptance test until it passes to the accept or reject zone. It may be hard to implement a 'reject' decision on the design but it must be done unless further improvements can be devised, in which case they also must be tested.

11.3 LAPLACE'S TEST FOR GROWTH/DETERIORATION

Laplace's test, which dates from 1773 making it one of the oldest in the statistical literature, is for the purpose of deciding whether a sequence of data collected over time exhibits a trend, i.e. a tendency to increase or decrease with time. Absence of trend, or stationarity may be defined as a lack of any long-term tendency to increase or decrease with time in the means of sample sequences of data taken at successive epochs in the system life. The means need not all come out the same, but there should be no definite progression. To be non-stationary on the other hand, $\theta(T_1) > (<) \theta(T_2)$ for all $T_1 < T_2$ would be too stringent a condition; there can be a tendency to increase or decrease without going that far. Laplace's test allows for some variations. It is based on the times from system new to each failure, i.e. $T_i, i = 1, \ldots, n$. If there is no trend, then the T_i are order statistics of a uniform distribution on $0, T_n$.

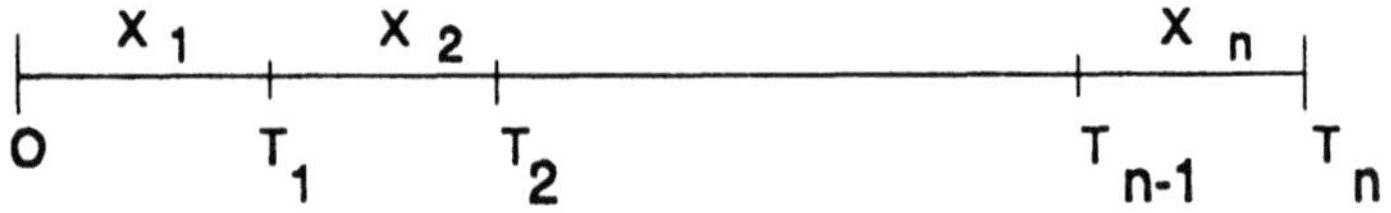

Fig. 11.4 Relationship between X_i and T_i.

It can be shown that approximately and for $n \geqslant 4$, the statistic U defined below has a standard normal distribution such that U is positive for an increasing tendency and negative for a decreasing tendency. The area under the normal curve, at U s.d.s, $\Phi(|U|)$, measures the confidence that the tendency is increasing or decreasing

$$U = -[12(n-1)]^{\frac{1}{2}}\left[\sum_{i=1}^{n-1} T_i/T_n(n-1) - \frac{1}{2}\right] \tag{11.23}$$

$$= -[12(n-1)]^{\frac{1}{2}}\left[\sum_{i=1}^{n-1}\left\{\sum_{j=1}^{i} X_i\right\} \Big/ (n-1)\sum_{i=1}^{n}(X_i) - \frac{1}{2}\right]. \tag{11.23a}$$

Note that the last entry, T_n is always omitted from the summation. Thus if U was found to be $+2.0$ we could be 97.7% confident of an increasing trend in $X_i = T_i - T_{i-1}$.

An alternative test, given in MIL-HDBK 189 (1981) has a test statistic V which is distributed as χ^2 with $2n - 2$ degrees of freedom, i.e.

$$V = 2\sum_{i=1}^{n} \log_e(T_n/T_i) = 2\sum_{i=1}^{n} \log_e\left[\sum_{i=1}^{n} X_i \Big/ \sum_{j=1}^{i} X_j\right]. \tag{11.24}$$

According to Ascher and Feingold (1984), the Laplace test is optimum if the ROCOF is increasing or decreasing exponentially, i.e.

$$\text{ROCOF}(T) = \exp(a + bT)$$

while the MIL-HDBK 189 test is optimal against a power function, i.e.

$$\text{ROCOF}(T) = aT^b.$$

Unfortunately, neither test is particularly effective at picking the bathtub curve, or even one which is constant for a longish period before increasing, unless the test is performed over intervals of the total time span taken in succession. In these circumstances the data may sometimes be too few to obtain decisive results.

A test for stationarity should always be applied before any further work which assumes that the MTBF is constant is done. The χ^2 test and the sequential test both involve this assumption and so before an acceptance test result is accepted, a Laplace or MIL-HDBK 189 test should be made on its data to ensure that the ROCOF is not increasing. (It does not matter so much if ROCOF is falling – in that case specifications may have been exceeded.)

11.4 RELIABILITY DETERIORATION – OVERHAUL PLANNING

Overhaul may be considered as a substitute for renewal which will be adopted if it is cheaper in the long run. Alternatively, it is a method of concentrating PM into short bursts so that the rate of expenditure between overhauls is lowered. The usual reason for doing this is that planned downtime between these periods is difficult to arrange and (or therefore) expensive.

Drinkwater and Hastings (1967) showed for a renew-at-first-tangent regime that if an asset suffered a major failure at T, the limiting cost for repairs should be $r(T)$, the difference between the total cost (purchase plus upkeep) curve and its tangent at the origin as shown on Fig. 11.5. $r(T)$ was also the value of an asset in average condition and with no extant failure at age T, that is $r(T)$ is a fair price at which to buy or sell it.

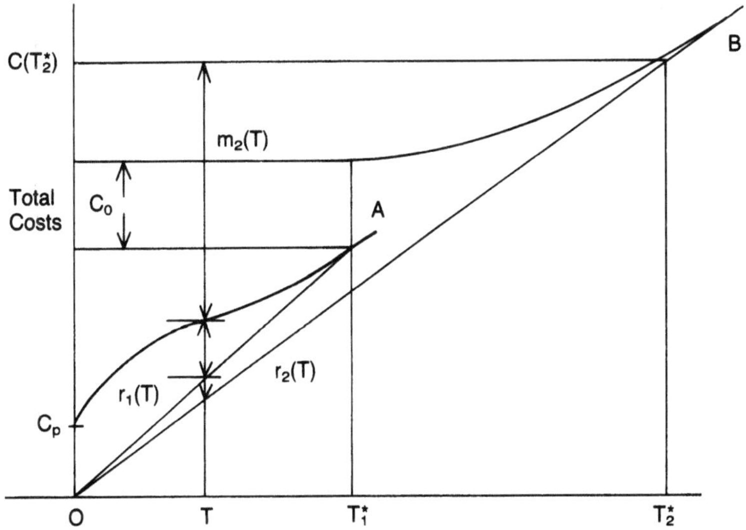

Fig. 11.5 Total cost curve showing overhaul as alternative to renewal.

This situation is not altered by overhaul, which is a form of re-investment short of renewal, unless the repair affects the cost of the next overhaul, which it may well do. Also bringing forward the overhaul to coincide with the repair might be worthwhile, if the overall saving was sufficient. The success of a maintenance regime can be judged by whether or not it results in a constant ROCOF over a reasonable working life. If equipment is overhauled then we may expect the ROCOF to rise between overhauls, slowly or not at all at first, and then increasingly rapidly, as the overhaul interval is increased. Many equipments are discarded when their ROCOFs start to increase (due to effects described in the section on the bathtub curve in Chapter 1), because not only does the rate increase, but the mean cost of the failures also rises, sometimes quite sharply. We define overhaul as an operation intended to restore the complex equipment to either good-as-new condition or to a predefined standard condition. The standard overhaul will result in a standard post-overhaul ROCOF pattern on average over a population of similar systems. Usually the work package for the overhaul is standardized and comprehensive rather than conditional upon what is found, so that equipments start the next phase in the same condition.

If the overhaul costs, the ROCOF pattern and the failure cost pattern are all known, then the overhaul interval can be optimized. Maintenance optimization as a whole is a subject for another book, but it is interesting to note that the problem of dividing work optimally between overhauls and intermediate routine PM has so far received little attention in the literature, probably because it is intractable without unreasonable assumptions. Nor has the question of how many overhauls an equipment can economically take before being scrapped been properly tackled. Reliability and costs interact in very complex ways which, together with design obsolescence, make it very difficult to calculate optimal policies. The situation is more complicated than many published models would have the reader believe.

Nevertheless, if the schedule of PM between overhauls is regarded as fixed, and a large number of replications of the item exist and are treated more or less equally, then it is possible to establish overhaul intervals based upon records of rising cost per unit time, i.e. an empirical solution based on acceptance of the existing PM schedule.

The figure indicates the following:

1. First overhaul or renewal should occur at the tangent to the total cost curve from new, since this minimizes the average cost per unit time.
2. Whether this should be a renewal or an overhaul depends upon the cost of the overhaul and the upkeep costs thereafter. If the tangent OB is steeper than the first tangent, OA, then renewal is indicated, otherwise we should overhaul. Alternatively, the PM policy should be adjusted to make the slopes equal.

3. If the overhaul cost C_o is greater than the purchase cost C_p it is unlikely but not impossible that overhaul will be cost effective because the cost-rate of ownership (gradient of the curve) after overhaul is likely to be steeper on average than that before the first overhaul. This factor will also make successive overhaul intervals shorter and limit the number of overhauls which can be justified economically.
4. The shapes of the curves depend upon the PM done between overhauls and before the first overhaul. This interdependence means that optimisation of the whole PM, overhaul and renewal scenario is very difficult.

If $AT_1^*/OT_1^* \geqslant BT_2^*/OT_2^*$ then overhaul will be worthwhile, and similarly for a second overhaul. The repair limits will also be altered if overhaul(s) are performed.

Referring to the figure:

$$\text{without overhaul} \quad \frac{C_F + m_1(T)}{T_1^* - T} < \frac{r_1(T) + m_1(T)}{T_1^* - T},$$

$$\text{with overhaul} \quad \frac{C_F + m_2(T)}{T_2^* - T} < \frac{r_2(T) + m_2(T)}{T_2^* - T},$$

where $r(T)$ is the repair limit and C_F the repair cost estimate. If $C_F < r(T)$ the decision is to repair otherwise renew or overhaul $r_2(T)$ will be greater than $r_1(T)$ if the gradient of the tangent at T_2^*(OB) is less than that at OA as shown. If the gradient of OA is greater than that of OB, then $r_2(T) > R_1(T)$. Thus, consideration of overhaul as an alternative to system renewal can result in considerable savings. It seems plausible, *prima facie*, that timing the overhauls so that the first and succeeding tangents OA, OB, . . . , were all coincident would be in some way optimal over other policies, but this remains to be proved.

REFERENCES

Ascher, H. and Feingold, H. (1984) *Repairable Systems Reliability: Modelling, Inference, Misconceptions and their Causes*, Marcel Dekker, Basel.

Cox, D.R. and Lewis, P.A. (1966) *The Statistical Analysis of Series of Events*, Methuen, London.

Crow, L.H. (1974) in *Reliability and Biometry* (eds F. Proschan and R.J. Serfling), Society for Industrially Applied Mathematics (SIAM), Philadelphia, pp. 379–410.

Crow, L.H. (1975) Tracking reliability growth, 20th Conference on Design of Experiments, 75-2, pp. 741–54.

Crow, L.H. (1983) *On Methods for Reliability Growth Assessment during Development*, NATO Advanced Science Institute on Electronic Systems Effectiveness and Life Cycle Costing, Springer-Verlag, NY.

Downton, F. (1969) *Journal of the Royal Statistical Society*, Series B, **31**, 335–49.

Drinkwater, R.W. and Hastings, N.A.J. (1967) *Operational Research Quarterly*, **18**(2), 121–38.
Duane, J.T. (1964) *IEEE Trans.*, **A-2**, 563–6.
Jardine, A.K.S. (1981) *Maintenance Replacement and Reliability*, Canadian Scholars Press, 211 Grenadier Rd, Toronto M6R 1R9 (earlier edition 1973 Pitman, London).
Lee, L. and Lee, S.K. (1978) *Technometrics*, **20**, 41–5.
Lloyd, D.K. and Lipow, M. (1964) *Reliability: Management Methods and Mathematics*, Prentice Hall, NJ.
MIL-HDBK 189 (regularly updated) *Reliability Growth Management*, HQ, US Army Communications Research and Development Command, Fort Monmouth, NJ 07702 (see criticisms in Ascher and Feingold, 1984).
MIL-HDBK 217 (regularly updated) *Reliability Prediction of Electronic Equipment*, RomAir Development Center, Griffiss Air Force Base, NY 13441.
MIL STD 781 (regularly updated) *Reliability Design Qualification and Production Acceptance Tests*, Naval Electronic Systems Command, Washington, DC, 20360.
Wald, A. (1947) *Sequential Analysis*, Wiley, NY.

CHAPTER 12

Data analysis

12.1 INTRODUCTION

Reliability analysis of systems cannot advance beyond theoretical formulae without data from real life or from tests to put some numbers in place of the symbols. This point will be further emphasized and developed in the next chapter. In this chapter we are concerned mainly with the mechanics of data analysis. We do not deny the importance of effective mechanisms for the collection of data, but we cover these only briefly. We explain what outputs are required from the data system rather than prescribing how it should be done because we feel that such methodology is more appropriate to a book on maintenance management and optimization.

We describe Pareto or modes analysis first because it is, in our view, a necessary preliminary to distribution analysis. Distribution analysis is sometimes indicative of the causes of failure but its main purpose is as an input to models for spares provision and for the optimization of intervals for component renewals in maintained complex items. In such cases availability or productiveness is the measure of merit and the bathtub curve effect usually means that all items suffer pseudo-Poisson processes and so may be taken to have constant ROCOFs and (almost) constant repair rates. However, in the case of unmaintained systems, reliability becomes the figure of merit and the distributions of failure times of individual components may be very important in assessing whether a system has met or is likely to meet a target value.

For example, if a component's failure time distribution is known to be normal with mean 500 h and s.d. 100 h, then its reliability over the first 200 h from new is the area under the standard normal curve from $\mu - 3\sigma$ to ∞ which is 0.998 65. If we knew only that the mean was 500 h and so were forced to assume an exponential distribution, the reliability would be estimated as $\exp(-200/500) = 0.6703$. This involves a factor of 244 in the unreliability or probability of failure! However, if the component under investigation was part of an item which was maintained by renewal of failed components and the age of this component was unknown, then the exponential assumption would be the only one we could make, even though the distribution function was known (Chapters 1, 2, 11). Thus a motor vehicle owner might renew rapidly wearing parts such as fanbelts, water hoses and brake pads before undertaking a long tour of faraway places, so that they would still be young and unlikely to fail when he returned from his holiday. Alternatively, he might accept the increased probability of inconvenience from failure, but take a spares pack along with him on the tour. An experienced reliability engineer would allow time between the parts renewals and setting out for the results of bad workmanship to become evident.

12.2 PARETO ANALYSIS

12.2.1 Principle

Pareto analysis is based upon the principle of the 'vital few and the trivial many' first extracted from the literature of economics as a 'universal truth' by Joseph Juran (1975) the great guru of quality. It is certainly common experience that failures to complex items are mostly (say 80–90%) in just a few (1, 2 or 3) modes of failure or components of the system. This is only to say that every design has its weak points. One might retort that these weak points should be designed out. But then the (empirically confirmed) theory of the Pareto principle holds that the elimination of the worst modes simply reduces the scale of the pattern without altering its general shape.

12.2.2 Setting up

In setting up a Pareto analysis, there may be a failure modes and effects analysis (FMEA) (Chapter 13) which delineates possible modes of failure. In that case, actual failures can be classified accordingly as they occur, always keeping in mind the possibility of additional modes being found. Also, some theoretically possible modes will perhaps not occur at all. If no FMEA exists then the greatest care must be taken in classification so that failures which are grouped together really were (near-) identical and not simply affecting the same components. This is because any subsequent design or maintenance schedule change or operational action must be specific to the causes of failure if it is to be fully effective.

12.2.3 Method

In the following example, which illustrates the mechanics of this method of analysis, one of us (DJS) was partly frustrated in his efforts to improve the maintenance of some acid pumps at a fertilizer factory by the maddeningly general and uninformative nature of the pre-existing classification system. This system did not record sufficient detail as to components affected and left no clue as to the causes of the failures. For example, were the 'holes' generally due to corrosion (most likely) or to mechanical

Table 12.1 Classification of failures of acid pumps (from DJS's PhD Thesis, Loughborough University of Technology, 1979)

Year	*1976*	*1978*
Blockages	49	32
Leaks	107	28
Drive	38	48
Electrical	28	20
Holes	156	54
Instruments	2	5
Valves	38	51
Other	14	25
Total	432	263

causes? What precisely constituted the dividing lines between electrical and instruments, or leaks and holes? Two sets of data were collected, for the years 1976 and 1978. During 1977, changes to the maintenance policy were implemented which reduced the total number of failures from 432 to 263 per annum. Because of the non-specificity of the classification system no 'mode' was actually eliminated but many were reduced, because even this crude classification, together with local knowledge, was enough to highlight where the major problems were. However, this remains an example of how not to classify failures.

The raw data were tabulated as in Table 12.1 which is an extract from a more comprehensive table covering other types of equipment as well.

12.2.4 Presentation

The usual method of presentation of the data (for 1976) is shown in Fig. 12.1. We see little point in the cumulative graph above the histogram and prefer to present the data as in Fig. 12.2 which has the added advantage of being able to show changes if required. The horizontal presentation is generally easier to generate as well as easier to read, in our view. Also, it presents all the information in one figure.

12.2.4 Other orderings

Ordering by frequency is the most common form but it is also possible to place the failures in order of total cost (i.e. number of failures in that mode × average cost). Other orderings such as according to unavailability caused by the mode, or lost production, or repair manhours, are also useful in certain contexts.

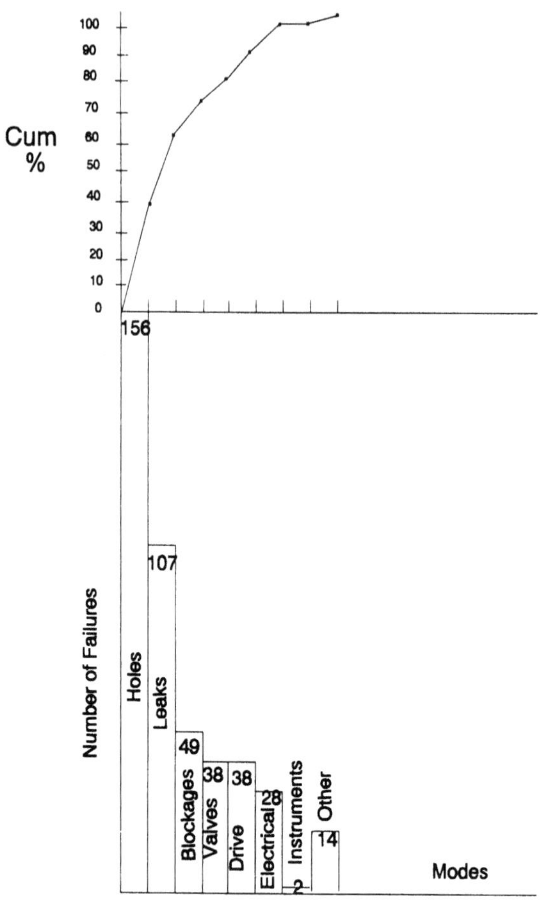

Fig. 12.1 Pareto analysis of 1976 data from Table 12.1 presented in the conventional manner used by Juran (1975).

12.3 DISTRIBUTION ANALYSIS

12.3.1 Modes

Pareto analysis is an essential preliminary to frequency or distribution analysis because failure times from different modes of failure cannot be expected to be independently and identically distributed (i.i.d.). If two or more modes are mixed up, then their combined distribution will also have two or more modes in the statistical sense. Even if the same component(s) is (are) involved in two modes, distinguished only by having different causes, this will occur. Methods are available for separating two or more modes' distributions statistically, but they are not particularly satisfactory. It is much better to be careful to separate the modes correctly during data collection, if possible.

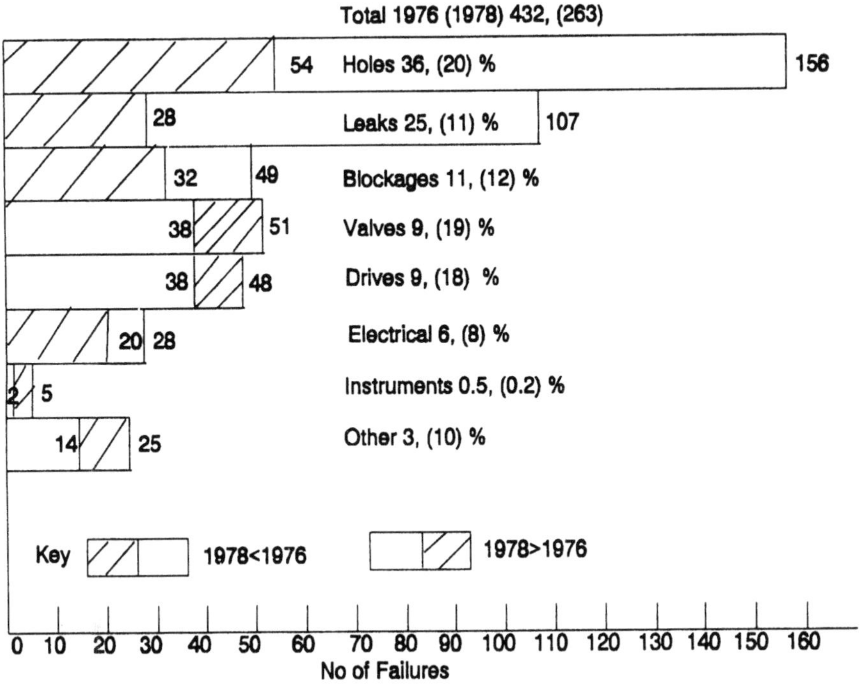

Fig. 12.2 Pareto analysis of 1976 and 1978 and pump data arranged for direct comparison and containing all information from Table 12.1 and Fig. 12.1.

12.3.2 Stationarity

In Chapter 11, we introduced readers to the Laplace and MIL-HDBK 189 tests for trends in failure-time data collected over a period of time from the same machine. In general we believe, from a combination of experience and common sense, that the considerations leading to the bathtub curve's constant ROCOF portion, namely the elimination of early life problems plus a constant maintenance policy, constant design configuration and constant conditions of stress in operation, are quite commonly achieved or approached in real life. If the ROCOF is constant for the whole machine under these conditions then it seems reasonable to assume that failure times for any particular mode of failure will be i.i.d. All the same, the tests are available and should be used to check that the MTBF is not changing over calendar time, that is that the sequence of failure times is not serially correlated with the overall age of the machine. It is, as Ascher and Feingold (1984) have pointed out at considerable length, potentially very dangerous as well as mathematically wrong to treat a non-stationary stochastic point process as though it were a stationary distribution. Besides the Laplace or MIL-HDBK-189 test it is often instructive to eyeball the sequence of failures on a calendar or hours-run-from-machine-new basis as in Fig. 12.3.

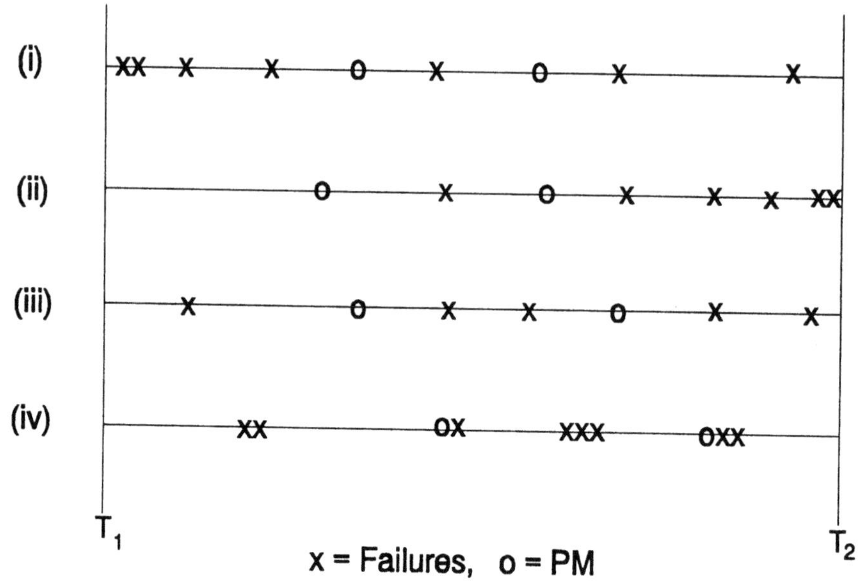

Fig. 12.3 Four patterns of calendar sequences of failures and PM for single machines: (i) improving reliability; (ii) deteriorating reliability; (iii) stationary MTBF; (iv) evidence of poor workmanship or spares quality.

In Fig. 12.3 (i) the failure times are clearly increasing. This sequence should not be analysed for distribution but if it were, it would probably be judged as having decreasing $z(t)$ because there are a number of very short times, some long ones and not much between. However, it is the ROCOF which is falling, not the hazard rate, and being non-stationary it cannot properly be said to have a hazard rate function at all. The sequence at (ii) is simply the reverse of the first sequence. By inspection, the MTBF is falling, the ROCOF increasing. Analysed (improperly) as a distribution, the results would be identical to (i), that is falling hazard rate. This sequence should not be analysed as a distribution because it is not stationary. Incidentally, its 'failure rate' is rising and falling at the same time – a good reason to abandon this indefinite term. Sequence (iii) shows no obvious increase or decrease in MTBF and would probably 'pass' the trend tests. Nor is there any tendency to fail shortly after PM or previous failure repair as there is in sequence (iv). Sequence (iii) can safely be analysed for distribution provided that it refers to a single mode of failure rather than the whole machine. The MTBF for a whole machine may well be stationary, but because it is a mixture of component failures it will almost certainly turn out to fit the exponential distribution because it is in fact a homogeneous pseudo-Poisson process (HPP). This was explained in more detail in Chapters 1 and 11. Distribution analysis usually will not tell us anything new for a whole machine. Sequence (iv), whether for a whole machine or a single mode, is likely to analyse for distribution either as decreasing $z(t)$ (hyperexponential) or as bimodal. It is really bimodal but if there are insufficient data this may not show up in the analysis. When the failures arrive in bunches (like London buses!) it is *prima facie* evidence of poor workmanship or poor-quality spare parts which should be investigated. It is most important that maintenance schedules are not predicated upon a false analysis of distribution. This was discussed at length by Sherwin and Lees (1980) in a paper describing data analysis in the chemical industry.

12.3.3 Renewal

Where failure times are effectively the times between renewals of components which have failed, there is an expectation that they will be truly i.i.d. However, even in these seemingly perfect cases it is possible that deterioration (upward trend in ROCOF) will occur if mating or otherwise influential components are not also renewed. Very often, first failures from new have a distinct distribution from all later failures which may be bimodal (early quality failures and regular failures or possibly original parts and replacement parts being of different qualities). In other cases repair consists of refurbishment and/or adjustment and the ROCOF may go either way or be reasonably constant. In general, if the repair returns the machine to a standardized condition then i.i.d. is likely but never certain. An i.i.d. assumption may be justified over a portion of machine total life for a given component but may suffer a change to a different but still valid distribution over another part of life. The null hypothesis of both the Laplace and MIL-HDBK-189 tests is that the figures are i.i.d., so we are pretty well certain (95% or 99%) there is a trend when a set of data fails either test.

In general we conclude that although there may be a few arcane exceptions, in reliability data analysis, i.i.d., renewal process and constant MTBF usually all mean much the same thing, namely a data set upon which it is safe to proceed to distributional analysis, provided that it can be stated with reasonable certainty that only one mode of failure is involved.

12.3.4 Test and field data

It is generally easier to analyse data from tests than from the field. Two factors are involved in this: first, in a test the investigator who is in charge, sees everything, and can design the test to suit the analysis requirements rather than *vice versa* as in analysis of field data; and second, the standard methods of analysis for distribution were developed originally by experimental scientists and engineers for their own use and need interpretation and modification for use with field data.

Data from tests are usually of only two kinds: complete times to failure and right-hand censored. This requires some explanation. A datum is said to be censored if either the start or the failure time is unknown. Usually it is the failure time that is missing because the test or the period of observation of the system terminated between two failures in the mode of interest. This is right-hand censoring. Left-hand censoring occurs in field data. When observation does not start from system new (and for bathtub curve reasons it probably should not do so), or from the repair of the component of interest, then there is a component life for which the end-time is known but not the beginning. Censoring (right-hand) also arises if components are renewed because of failures in a mode other than the one being analysed, or as part of the preventive maintenance plan.

It is important to realize that censored lives are evidence of reliability, just as failures are evidence of unreliability. It is incorrect to ignore them; they must be included in the analysis. All the information that can be extracted from the data is relevant to decisions based upon the records.

A generalized and typical field data set would involve several, say N, nominally identical items operated in a similar manner over a 'window' of calendar time as in Fig. 12.3 with records kept of failures and preventive renewals of the component (or mode of failure) in question. It would thus contain N left and N right censorings at either end of N separate series of events. Each of these series should strictly be subjected separately to tests for serial correlation, but often the sample size is inadequate for conclusive results. Also, before combining them into one data set, one should really check the variation in the means and variances of the sub-samples, but again the samples are usually too small to make such a procedure convincing. Nevertheless, it is important to be reasonably sure that all the data are from one distribution. If they are not, then the bi- (or multi-) modality may show up in the detailed analysis, or then again it may not. If it does not, then policies may be applied to all examples of the similar equipment when it would be better to have different policies for individual machines according to their use. On the whole, though, if the designs are identical and the treatment similar, a single distribution is to be expected.

Carter (1972, 1988) has given two examples to illustrate the care which must be taken. First, a petrol engine used in a luxury car where it was hardly ever stressed beyond 20% of its capacity was fitted in an armoured vehicle where it had to work much harder. The failure-time distributions for several wearing components were quite different in the two applications. The second example illustrates an unexpected danger. It was found that the variance of times to failure of tank manual gearboxes was unexpectedly wide. Investigation showed that the variation was mostly due to differences in skill between drivers. In a later model of tank, automatic gearboxes were fitted and the failure times were definitely all of one distribution.

12.3.5 Development of analytical methodology

We shall explain the methodology of distribution analysis in a progressive way, starting with the analysis of complete, uncensored test results and progressing in steps to the analysis of generalized data such as that arising from use of equipment in its intended role. Experience in teaching this topic shows that it is very necessary to point out the difficulties and pitfalls first rather than afterwards. Otherwise a complacent attitude is engendered that reliability data analysis is easy, which it is not.

12.3.6 Choice of distribution form

In Chapter 1, a number of distribution forms were discussed. The Weibull was shown to be a very versatile form able to represent a wide range of distributions. Because of this versatility it is usual to analyse reliability data first on the assumption that the distribution is of Weibull form. If it does not plot out well then the aberrations will, with experience suggest what other forms of distribution should be tried.

12.3.7 Analytical methods

Methods exist for estimating the parameters of distributions from data by analytical means. However, we consider the graphical methods described below to be more useful practically. The graph shows the experienced reliability engineer a lot more than a printout of maximum likelihood parameter estimates. Analytical methods are generally asymptotic; accuracy with them is not achieved without lots of data. Graphical methods permit engineering judgement to override mathematics, which has its own dangers but which is generally better overall for the small data sets (10–50 data), that are usual in reliability work. For more details of analytical methods, readers are referred to the excellent book by Mann, Schafer and Singpurwalla (1974).

12.3.8 Weibull analysis (two-parameter) of uncensored test data

This is the simplest analysis, but it is the basis of the methods which cover censored data and three-parameter distributions.

Recall from Chapter 1 the form of the two-parameter Weibull cumulative distribution function

$$F(t) = 1 - \exp[-(t/\eta)^{\beta}] \tag{12.1}$$

take logarithms to the base e twice to obtain

$$\log_e \log_e[1/(1 - F(t))] = \beta \log_e t - \beta \log_e \eta. \tag{12.2}$$

Weibull probability paper uses a log-log reciprocal vertical scale and a logarithmic horizontal scale so that $F(t)$ for the two-parameter Weibull plots as a straight line against t. This straight line has slope β. The scale parameter η can be determined by considering the point at which $t = \eta$. Regardless of the value of β, when $t = \eta$, $F(t) = 1 - 1/e = 0.6321$.

The data in this simple analysis are all times to failure. Let their number be n, and the times to (or between) failures be labelled t_i, $i = 1, \ldots, n$, where t_1 is the smallest, $t_i \geqslant t_{i-1}$ and t_n is the largest. In order to make a plot from which to estimate β and η, we need to decide what value, $F(t_{i/n})$, of $F(t)$ to assign to the ith failure out of n. The number of failures out of n that occur before time t has a binomial distribution with mean $nF(t)$. This is so whatever the form of $F(t)$. The probability of exactly i failures before t is

$$Pr(i,t) = \binom{n}{i} F(t)^i [1 - F(t)]^{n-i}.$$

The binomial is related to the beta distribution by the identity

$$Pr[(B:n,p) \geqslant i] \equiv Pr[(\beta:i, n - i + 1) \leqslant p]. \tag{12.3}$$

Substituting $F(t_{i/n})$ for p in equation (12.3) above tells us that the likelihood function (or p.d.f.) on the $F(t)$ scale of the ith failure out of

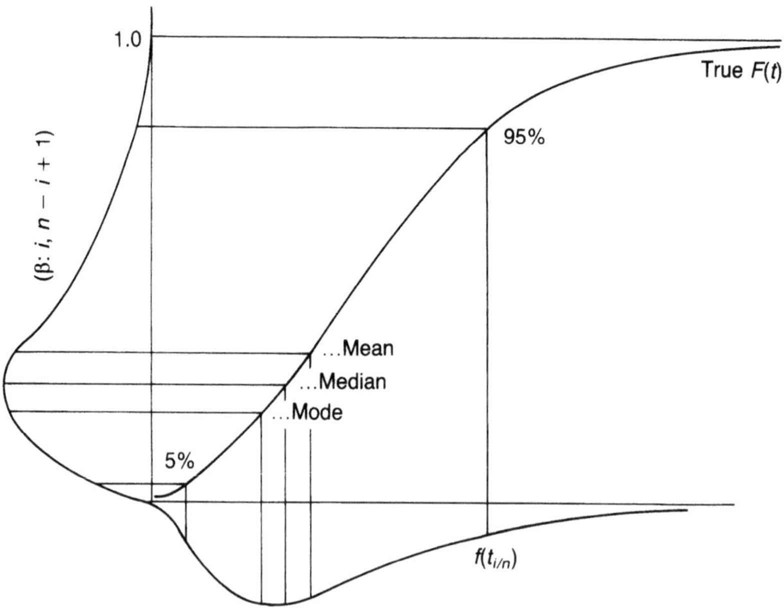

Fig. 12.4 Relationship between p.d.f. of times to ith-ordered failure out of n in repeated trials $f(t_{i/n})$, true c.d.f. $F(t)$ and the likelihood function of $F(t_{i/n})$ of beta form with parameters i and $n-i+1$.

n (in repeated trials of scale n) is of beta form with parameters i, and $n - i + 1$.

Figure 12.4 shows a graph of the true $F(t)$ function on linear scales and its relationship to the p.d.f. of ith failures out of n in successive trials. The reflection of this p.d.f. in the $F(t)$ given on the $F(t)$ scale is always the same for the same i and n whatever the form of $F(t)$. The figure shows the mean, median and mode corresponding between $f(t_{i/n})$ and $\beta : i, n - i + 1$, but any points of equal probability can be constructed in the same way.

The natural estimator of $F(t_{i/n})$, namely i/n, is easily shown to be unsuitable. Consider the last point $n/n = 1$. The data set is a sample but this estimate implies that no failure time greater than t_n is possible ever, which is clearly false. Again, the instinctive $F(t)$ estimate for the fifth out of nine is 50% since there are four failure times greater and four smaller, but 5/9 = 55.5%. We may conclude that i/n is an overestimate for $F(t_{i/n})$. Figure 12.4 shows the three measures of central tendency mean, median and mode, most commonly used in statistics. These represent respectively the value that is right on average, the value that is too high in 50% of cases (and too low in the other 50%) and the most likely value. In the beta distribution with parameters i, $n - i + 1$, these take the following values

$$\begin{aligned} \text{mean} &= i/(n+1) \\ \text{median} &\approx (i - 0.3)/(n + 0.4) \text{ (Bénard's estimate)} \\ \text{mode} &= (i - 1)/(n - 1),\ i > 1. \end{aligned} \tag{12.4}$$

We note that whichever way the distribution is skewed, the median is always between the mode and the mean.

The mode is seldom used, principally because the first point is always lost. The mean rank $i/(n + 1)$ is very easily calculated and is popular if n is large. For small samples it is usual to use the median rank. Bénard's estimate is very good for $n > 5$, but 50% and other percentage ranks can only be calculated properly by solving equations in the form of (12.5) below for the 100 P% rank of the ith failure out of n, where R_p is the required rank

$$P = \binom{n}{i} R_p^i (1 - R_p)^{n-i}. \tag{12.5}$$

Equation (12.5) is difficult to solve, so usually the ranks are found from the c.d.f. of the beta distribution, directly by numerical integration, i.e.

$$R_p = \int_0^p x^{i-1}(1 - x)^{n-i}\mathrm{d}x \Big/ \int_0^1 x^{i-1}(1 - x)^{n-i}\mathrm{d}x. \tag{12.6}$$

The denominator of equation (12.6) is known as the beta function and is most easily computed from

$$\beta(v,w) = \Gamma v \Gamma w / \Gamma(v + w) \tag{12.7}$$

$$= (v - 1)!(w - 1)!/(v + w - 2)!$$

$$\text{for } v \text{ and } w \text{ integer.} \tag{12.7a}$$

If v and w are not integer, equation (12.7) can be solved using a polynomial approximation for the gamma function due to Stirling

$$\Gamma x = 1 - 0.577\,191\,652x + 0.988\,205\,891x^2 - 0.897\,056\,937x^3 + 0.918\,206\,867x^4 - 0.756\,704\,078x^5 + 0.482\,199\,394x^6 - 0.193\,527\,818x^7 + 0.035\,868\,343x^8. \tag{12.8}$$

The gamma function is required to lie between 1 and 2 using

$$\Gamma x = (x - 1)\Gamma(x - 1), \quad x > 1. \tag{12.9}$$

A table of the gamma function derived from equation (12.8) is at the back of the book.

A method has been developed for plotting the times of n failures in a complete test (no censorings) using mean or median ranks in order to estimate the Weibull parameters β and η. Suitable graph papers, which plot the two-parameter Weibull as a straight line are made by Chartwell and TEAM among others. We have made our own in order to avoid copyright problems. With modern computer graphics and plotters it is not difficult to perform the whole analysis in the computer in the same way as it would be done by hand, thereby overcoming the disadvantages of previous non-graphical analytical methods.

EXAMPLE 12.1

The data in Table 12.2 refer to a test of 20 identical components fitted in a certain design of small switch. In this case the time variable is replaced by the number of on-off cycles to failure. The median ranks are taken from the table at the end of the book. (This table was developed from a computer program based on Simpson's rule and equation (12.6).) It was checked against similar tables given by O'Connor (1981, 1985) and others and found to agree to three decimal places. As two decimal places is about the limit to which the graph can be plotted, only two are given.

Another computer program was used to plot the data using a full-screen plotting device. This program performs the analysis in the same way as it would be done by hand, the points are plotted and the operation can shift and angle the line to obtain the best fit by eye.

The example data actually consist of random numbers from the distribution with $\beta = 1.62$, $\eta = 10\,235$. In this format, β is the actual gradient of the line. Readers should be aware on some commercial graph papers, the gradient is proportional to β but not equal to β, necessitating special geometrical constructions to obtain the estimate. The 63.2% line for estimating η is available on commercial papers. In the program used

Table 12.2 Data for switch example with median ranks

Failure no. i	*No. of operations* t	*Median rank* F(t) (%)
1	1290	3.41
2	2559	8.26
3	3055	13.1
4	3468	18.0
5	4465	23.0
6	5430	27.9
7	5800	32.8
8	6397	37.7
9	7392	42.6
10	7920	47.5
11	8530	52.5
12	9638	57.4
13	9751	62.3
14	10519	67.2
15	11514	72.1
16	12990	77.0
17	14262	81.9
18	15835	86.8
19	17995	91.7
20	21700	96.6

here, η and β are calculated from the line directly and printed out as shown in Fig. 12.5.

12.3.9 Single censoring

A data set is said to be singly censored if all the censorings are right-hand and all are greater than or equal to the longest failure time. This situation arises frequently when tests of scale n are terminated at m failures. That is, n items are placed on test at the same time, and the test is stopped when $m < n$ have failed. In this case the graph simply stops short at the mth failure. We have no knowledge of what happens after that point. The same pattern (distribution) may continue or another mode of failure might take over the dominant role.

Extrapolation is dangerous and should only be undertaken if engineering considerations make a change of distribution unlikely. On the other hand, if the test was only to show that less than 5% were expected to fail before a particular time, then censoring in this way is justified.

Statisticians talk of Type I (single) censoring for a test which is stopped after the mth failure and Type II for a test which is stopped at a predetermined time regardless of how many failures have occurred. The distinction is not important for analysis by this method.

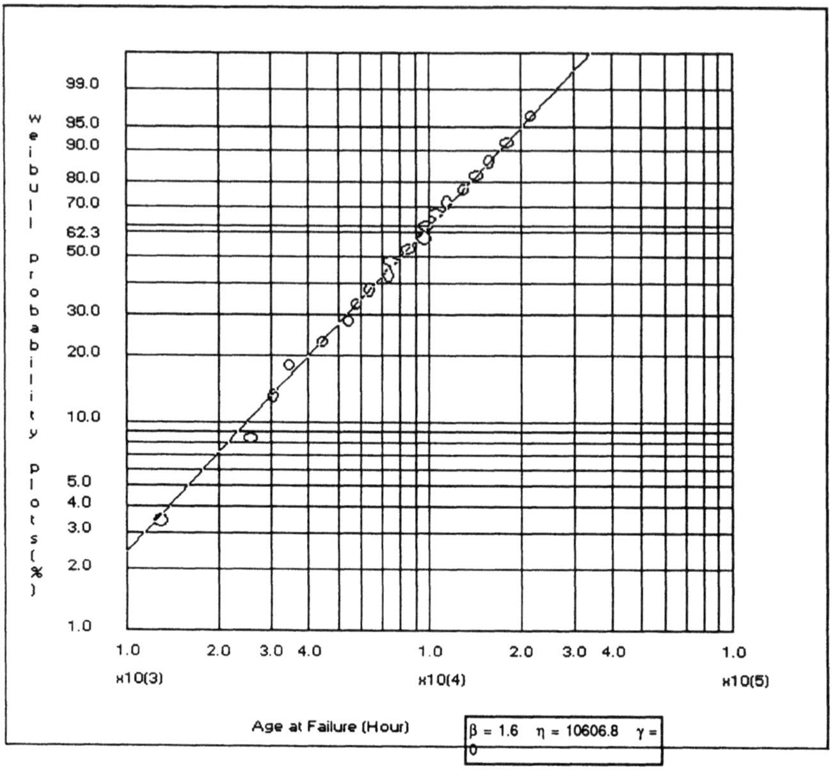

Fig. 12.5 Weibull plot of data in Table 12.2.

12.3.10 Multiple censoring

If some or all of the censoring takes place at running times less than that of the longest failure time, the sample is said to be multiply censored. The method of mean order numbers given below is able to cope with right-hand censoring of this type, which is sometimes described as Type III censoring, but not with left-hand censoring.

12.3.11 Mean order

A method is needed whereby proper weight can be given to lives which end in failures of a kind not being analysed or through other circumstances, notably preventive maintenance, and which can be used for any distribution form. With no censoring, as we have seen, the failure times are easily ordered from shortest to longest $(1 \ldots i \ldots n)$. The problem reduces to one of finding equivalent order numbers for the failures which do occur in the presence of censored lives. The assumption underlying the calculation of mean order numbers is that the failures prevented by censoring might have occurred in any of the intervals between subsequent events (failures and further censorings), or after all of them, with equal probability attaching to each interval. This assumption will be correct *on average*, (hence '*mean* order numbers'), given *no further information* about the censorings. It is *not valid* if the censorings occur other than randomly. In particular, it is invalid if items are withdrawn because condition monitoring shows that they are about to fail, but it is valid if items are removed at a certain age regardless of condition.

The mathematical expression for mean order numbers M_i for the ith failure in a set of censored data of scale n, i.e. a *total of n failures and censorings* is

$$M_i = M_{i-1} + \frac{n - M_{i-1} + 1}{1 - S_i}, \tag{12.10}$$

where S_i is the number of survivors just before the ith failure and includes the ith failure and any censorings simultaneous with it. The second term or increment remains the same until another censoring is passed.

The mean order number M_i in a censored sample analysis takes the place of i in a complete sample analysis. Linear interpolation between rank table entries is sufficiently accurate for engineering purposes. Note that the number of points for the graph is the number of failures. The order indicator i in M_i refers to the ith ordered failure time ignoring the censorings. In a computerized analysis equations (12.6) to (12.9) can be used with M_i substituted for i.

12.3.12 Unknown starts

In data taken from actual service over a calendar time window there are lives censored at their beginning (left-hand censorings). There is no satisfactory graphical way of dealing with the latter and they are best ignored except in calculating the best estimate of the mean time to failure. Although ignoring them discards information, the effect is simply to reduce the number of lives sampled.

Cox and Lewis (1966) show that the p.d.f. of the forward recurrence times for a stationary point process, (i.e. one that can legitimately be described by a distribution function) is given by

$$f_{fr}(t) = R(t)/\theta, \tag{12.11}$$

where θ is the distribution mean and $R(t)$ the reliability to time t. The forward recurrence time is the time from an arbitrary point T_0 in the point process ($T_0 \neq 0$) forward to the next event (failure). Equation (12.11) could be used in maximum likelihood calculations leading to parameter estimates, but cannot be incorporated into the graphical methods described here.

Clearly the best estimate of the distribution mean, θ is the total time at risk of all the equipments involved over the time window divided by the total number of failures occurring in the same period. Another estimate of the mean can be obtained from the parameter estimates $\hat{\beta}$ and $\hat{\eta}$ thus

$$\hat{\theta} = \hat{\eta}\Gamma(1 + 1/\hat{\beta}). \tag{12.12}$$

The two estimates should correspond more closely as n increases and should always be compared with a view to adjusting the parameter estimates.

12.3.13 Confidence limits of $F(t)$

By plotting the 5% and 95% ranks at the same times t_i as the median rank estimates, two curved lines are obtained which represent the confidence limits of $F(t)$. They approach each other most nearly across a common normal at $t = \eta$. The width of the confidence interval generally depends upon the number of data n. It is salutary to draw them in because otherwise it is possible to gain a false impression of accuracy from the often very close fit of the estimating straight line to the data points. Most reliability work is perforce done with rather small samples. With very high probability the true distribution lies somewhere within the two curved lines (5% and 95%), but there is even a faint possibility of it being outside them.

Some books advise plotting the limits in a different way which gives sets of points in horizontal rows at the same $F(t_i)$ value rather than in vertical columns at the same t_i value. We believe this to be wrong. To obtain smooth curves the limits should be plotted above and below the point on the estimated line corresponding to the median rank rather than above and the below the plotted point at t_i itself. Confidence intervals should be plotted if $n < 30$ and median (i.e. 50%) ranks should be used for the central line. For $n > 30$ it is sufficient to plot just the mean (or median) ranks.

12.3.14 Location parameter > 0

The Weibull distribution sometimes has a location parameter denoted by c or γ which modifies the distribution equations by substituting $(t - \gamma)$ for t. Thus, for example, the reliability becomes

$$R(t) = \exp[-\{(t - \gamma)/\eta\}^{\beta}]. \tag{12.13}$$

On the graph paper $F(t)$ for $\gamma \neq 0$ will plot as a curved line which becomes asymptotic to a line of slope β at the top and vertical at the bottom of its range. The value of γ may be estimated by several means. The easiest is to extend the curve downwards by eye until it becomes vertical and read off the value on the t-axis. A mathematical construction

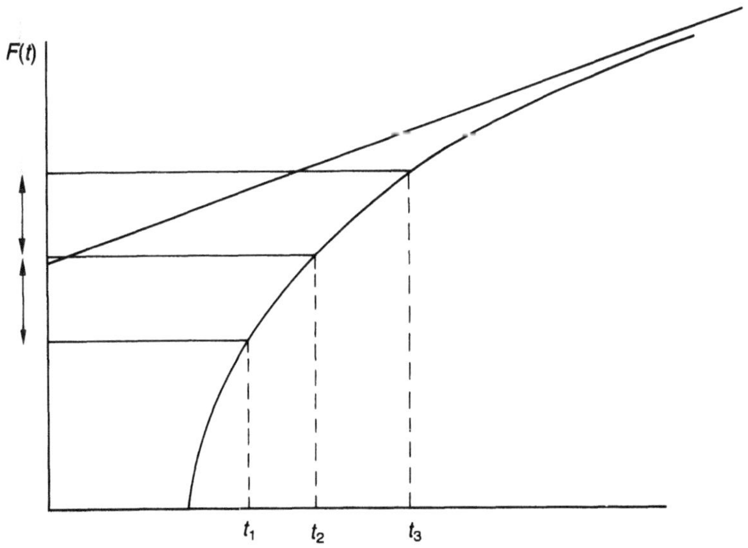

Fig. 12.6 Construction to estimate Weibull location parameter.

is to take three points on the curve $t_1 < t_2 < t_3$ such that the actual vertical distances from $F(t_1)$ to $F(t_2)$ and $F(t_2)$ to $F(t_3)$ are equal, then

$$\hat{\gamma} \approx t_2 - (t_3 - t_2)(t_2 - t_1)/(t_3 - 2t_2 + t_1). \qquad (12.14)$$

12.3.15 Bimodality

If there are two (or more) distributions present in the data and their means are sufficiently separated, the plot will have bends or breaks in it. If there are many distributions mixed up, as will occur when data from complex machines are not separated into nodes, then the result is likely to be $\beta = 1$ unless there is a dominant mode. The latter is a false analysis, $\beta = 1$ implies the exponential distribution whereas what we really have is a pseudo-Poisson process rather than a stationary renewal process. This is explained in more detail in Chapter 11. However, a bimodal appearance may occur if one mode of failure is dominant, especially if this mode has β considerably greater than 1. As stated earlier, it is better to separate the modes on engineering rather than statistical grounds, but if there is a dog-leg or a break in the plot and nothing further is known then it is legitimate to assume that (almost) all the failures before the change are of one distribution and those after it of another.

It is not sufficient, as some earlier authors have suggested, simply to measure the parameters of the two (or more) parts of the plot directly. Two (or more) new plots must be made after deciding to which mode to assign each of the failures. In each of these further plots the failures in the other mode (s) are treated as censorings. If bimodality is not recognized, as may be the case if the data are insufficient in number, then the shape parameter may well appear to be less than unity. An early mode due to random events rather than wearout is quite common in field data, and if the wearout is not pronounced, say $1 < \beta < 2$, then this is quite a likely outcome. A distribution which appears to have a negative location parameter is much more likely to be bimodal, since failure before operation is usually not possible.

12.3.16 Other distributions

The method of analysis explained above is applicable to any continuous distribution form. Special graph papers exist for the normal, lognormal and extreme value distributions as well as the Weibull. These are usually sufficient. The authors' advice is to analyse always graphically and for the Weibull distribution first. With experience, the appearance of the Weibull plot will suggest what other distributions may fit better. The following observations may be of use to beginners.

If $1 < \beta < 3$ the lognormal distribution, which is characteristic of metal fatigue failures may fit better. If $\beta > 3$ the normal, which is characteristic of abrasive wear, is often easier because it has only two parameters. However the mean θ needs to be greater than three times the s.d., σ.

The extreme value or Gumbel distribution of the largest extreme is not of much use – a Weibull will usually fit the data as well – but the smallest extreme version is able to represent negative skew. Such a distribution analysed as Weibull, will appear to have increasing β towards the top but otherwise fit quite well, but without the characteristic sharp dog-leg of bimodality. The smallest extreme value distribution is characteristic of the processes where item failure is from the first of many possible, similar failures to occur, such as the first of many gear teeth or turbine blades in a wheel to crack or the first corrosion pit on a pipe to become a hole.

12.3.17 Cumulative hazard analysis

This is another method of estimating the Weibull parameters. It is not suitable for any other distribution except the extreme value. It is particularly designed for grouped data in the form: 0–50 h, three failures; 50–100 h, four failures; . . . ; 400–500 h, two failures. It can be used for single failure data, but the more usual form given above is more accurate.

Recall from Chapter 1 that the integral of the hazard rate function $z(t)$ is known as the cumulative hazard function $H(t)$. For the Weibull distribution with two parameters the cumulative hazard function is

$$H(t) = (t/\eta)^{\beta}. \qquad (12.15)$$

Taking logs of equation (12.15) we get

$$\log_e H(t) = \beta \log_e t - \beta \log_e \eta. \tag{12.16}$$

Therefore if we plot $H(t)$ versus t on log-log graph paper, and the distribution is two-parameter Weibull, the result will be a straight line of slope β. Also from equation (12.15), when $t = \eta$, $H(t) = 1$. It remains only to find a way to estimate $H(t)$ from the data. By definition, the mean hazard over a period t_i to t_{i+1} is the number of failures in the period divided by the average number at risk. Estimates of cumulative hazard are obtained by summing the hazard estimates for successive periods. By convention, the points are plotted at the start of each interval and the denominator is the number at risk at that time. The first point is at the time of the first failure since zero is not possible. An alternative convention is to plot at mid-interval taking the average number at risk. In the example below we use the first convention. The method deals with censored data simply by reducing the number at risk when calculating the hazard z for the interval. Location parameters greater than zero and bimodality can be dealt with in exactly the same way in $H(t)$ as in $F(t)$ plotting.

12.3.18 Example of cumulative hazard calculations

Table 12.3 shows data arising in the field from truck engine water pumps (data courtesy of D.W. Newton).

Table 12.3 Data for cumulative hazard example (Courtesy D.W. Newton)

Mileage (thousands)	*1(a) Failures*	*1(b) Censorings*	*2 Survivors (S)*	*3 Hazard (z)*	*4 Cum. haz. (H)*
0–5	0	1	68	0	0
5–10	2	4	67	0.0298	0.0298
10–15	3	3	61	0.0492	0.0790
15–20	2	2	55	0.0364	0.1154
20–25	1	1	50	0.0200	0.1354
25–30	3	1	48	0.0625	0.1978
30–35	3	0	44	0.0682	0.2661
35–40	1	3	41	0.0244	0.2905
40–45	1	7	37	0.0270	0.3175
45–50	0	2	29	0	–
50–55	1	4	27	0.0370	0.3545
55–60	1	7	22	0.0455	0.3999
60–65	0	6	14	0	–
65–70	0	3	8	0	–
70–75	2	1	5	0.0400	0.7999
75–80	0	1	2	0	–
80–85	0	1	1	0	–
>85	0	0			
Total	20	48			

The procedure to produce the table is as follows:

1. Col. 1. Write down *all* the events in increasing order, irrespective of whether they are failure times or censoring times.
2. Col. 2. List the values of S, the number of survivors immediately prior to each event (or, in the case of grouped data, at the beginning of the class interval). S_1 is the total number of items n. S_2 is n less the number of failures and censorings in the previous interval and so on.
3. Col. 3. The sample hazard z (i.e. the number of failures in each class interval, divided by the corresponding S) is listed.
4. Col. 4. The sample cumulative hazard H (which is simply the cumulative sum of col. 3) is listed.

H is an estimate of the cumulative hazard function $H(t)$ which is the integral of $z(t)$ from 0 to t.

12.3.19 Extreme value distribution cumulative hazard

The extreme value distribution can also be modelled by an $H(t)$ plot. Recall from Chapter 1 that for the EV distribution the hazard rate $z(t)$ is an exponential function of t.

$$z(t) = (1/b)\exp[(t - a)/b]. \qquad (12.17)$$

Integration gives

$$H(t) = \exp[(t - a)/b]. \qquad (12.18)$$

Taking logs to base e

$$\log_e H(t) = (t - a)/b. \qquad (12.19)$$

So, on log–linear paper $H(t)$ plots as a straight line of slope proportional to $1/b$. Also, from equation (12.18), $H(t) = 1$ at $t = a$. Having found a, the simplest way to find b is to invert equation (12.18) for any other point on the line. For example, when $t - a = b$ then $H(t) = \mathrm{e} = 2.718$ or alternatively when $t = 0$ then $H(t) = \exp(-a/b)$ provide simple points.

12.3.20 Comments on the cumulative hazard method of analysis

Several comments and caveats must be made about cumulative hazard plotting.

1. It was devised before the days of cheap computing to deal with large data sets quickly.
2. It is unwise to throw away knowledge of individual failure times and censorings in order to group data. There is an inevitable and now unnecessary loss of accuracy.
3. There is no direct way to produce an $H(t)$ plot with confidence limits, so it should not be used for small data sets.

Table 12.4 Comparison of rank estimates

Order no. i	$F(t_i)$ *rank %, n = 10* *Median*	*Mean*	*Mode*	*From* H(t)
2	16.2	18.2	11.1	19.0
9	83.8	81.8	88.9	85.5

4. Some authorities advise the use of $H(t)$ calculations to estimate $F(t)$ using

$$F(t) = 1 - \exp[-H(t)]. \qquad (12.20)$$

But consider the estimates for the second and ninth failures out of 10 in a complete test at Table 12.4. Notice that the estimates from $H(t)$ lie outside the range between the mean and modal ranks, and will therefore tend to under-estimate β. One reason to prefer median to mean ranks is that the median is always between the mode and the mean.

Also, whereas the other methods give symmetrical ranks, e.g. in Table 12.4 100 − 81.8 = 18.2, those from $H(t)$ are asymmetric. Given the binomial basis for ranks discussed above, they should be symmetrical about their middle values (s).

12.4 EXERCISES

1. Produce a Weibull probability plot for the water pump data. Estimate β, η, the mean life and the standard deviation of the life.

2. Listed below are 100 random Weibull failure times. Select 10 of these and construct a Weibull plot. Compare your result with those of others taking different samples of 10 times.

1955	863	1183	1189	2346	1278	1880	1515	1453	363
416	241	1305	802	966	443	763	1398	1643	396
1446	1170	1421	948	1070	386	1331	2375	1815	1302
1839	625	2043	1316	855	583	646	945	641	1208
208	968	1526	1892	1446	806	383	1583	1025	1524
405	778	736	607	457	1690	653	1894	734	1885
1167	2288	783	1104	1678	1029	386	1107	1528	1385
904	2147	2326	934	327	830	1023	1542	195	2262
2038	519	4066	2288	571	754	158	1365	1181	1072
490	489	694	1321	2588	586	1588	1373	202	933

3. Replot your data of question 2 on lognormal paper.

4. Replot your question 2 data on Weibull paper regarding as censored:

(a) the three longest times;
(b) the last three times in order of choosing them from the matrix of numbers;

Plot 5% and 95% confidence limits of the distribution function estimate.

REFERENCES

Ascher, H. and Feingold, H. (1984) *Repairable Systems Reliability: Modelling, Inference, Misconceptions and their Causes*, Marcel Dekker, Basel.

Carter, A.D.S. (1972, 1988) *Mechanical Reliability*, Macmillan, London.

Cox, D.R. and Lewis, P.A.W. (1966) *The Statistical Analysis of Series of Events*, Chapman & Hall, London.

Juran, J.M. (1975) Quality *Progress*, May, 8–9.

Mann, N.R., Schafer, R.E. and Singparwalla, N.D. (1974) *Methods for Statistical Analysis of Reliability and Life Data*, Wiley, NY.

O'Connor, P.D.T. (1981, 1985) *Practical Reliability Engineering*, Wiley.

Sherwin, D.J. and Lees, F.P. (1980) *Proc. I. Mech. E.*, **194**, 29.

CHAPTER 13

The management of system reliability

The majority of this book consists of methods of probabilistic or statistical calculation to obtain numerical answers in the form of system reliability, availability, etc. As with other branches and specialisms in engineering, however, calculation is important, but not the whole story by any means. In fact many would claim, and we would not disagree, that the practice of reliability engineering is ~80% management and ~20% calculation whereas in expounding its theory the percentages must be reversed. We incline to the view that a reasonably good job can be made of ensuring that a product or system is adequately reliable, etc., without many calculations. The sophisticated probabilistic and statistical methods which dominate this book and others provide a very necessary edge in a competitive world. There is no doubt also that the existence of data from which figures can be derived for the reliability factors considerably enhances the value of managerially based reliability studies.

In this chapter we describe and illustrate the use of the standard techniques of reliability engineering with reference to the so-called product cycle. One of us (DJS) sat on the BSI Committee which drafted BS5760, *Reliability of Systems, Equipments and Components*. Part 1 of this standard is entitled *Guide to Reliability Programme Management* and contains much useful advice concerning the reliability related management actions necessary for a successful product. This short chapter does not attempt to reproduce all the detail of the Standard Guide, but should be read in conjunction with it in order to enhance understanding. Chapter 11 on maintained system reliability is particularly relevant to the application of the principles contained in this chapter and in the Standard Guide, but all the other chapters of the book will find application, if not in one project then in the next.

13.1 DEVELOPMENT OF RELIABLE DESIGNS

BS:5760 *Part 1* (*Guide to Reliability Programme Management*) divides projects into four phases:

Table 13.1 Program activities during the principal phases of a project adapted from BS5760 Part 1 Table 1 by kind permission of the Director General of the BSI

Phase	*Activities*
Definition	Reliability feasibility study Statement of reliability objectives and requirements Reliability specification and contract formulation
Design and development (including initial manufacture)	Analysis of parts, materials and processes Analysis of established and novel features Failure modes, effects and criticality analysis Incident sequence analysis (fault tree analysis) Stress and worst-case analysis Redundancy analysis Reliability assessment Human factors Design change control Design review Design audit Safety programme Maintainability programme Parts and sub-assembly testing Performance testing Environmental testing Accelerated testing Endurance testing Reliability demonstration Data collection, analysis and feedback
Production	Preservation of reliability achievement Quality conformance verification Screening (run-in, bed-in or burn-in) of components and assemblies Reliability demonstration
Function and maintenance	Data collection, analysis and feedback Redesign/modification Maintenance

(i) definition
(ii) design and development
(iii) production
(iv) function and maintenance,

each of which is further divided into activities, some of which are iterated (see Table 13.1 which is taken from Table 1 of BS5760, Part 1). This is the basis of any program, but circumstances dictate emphases.

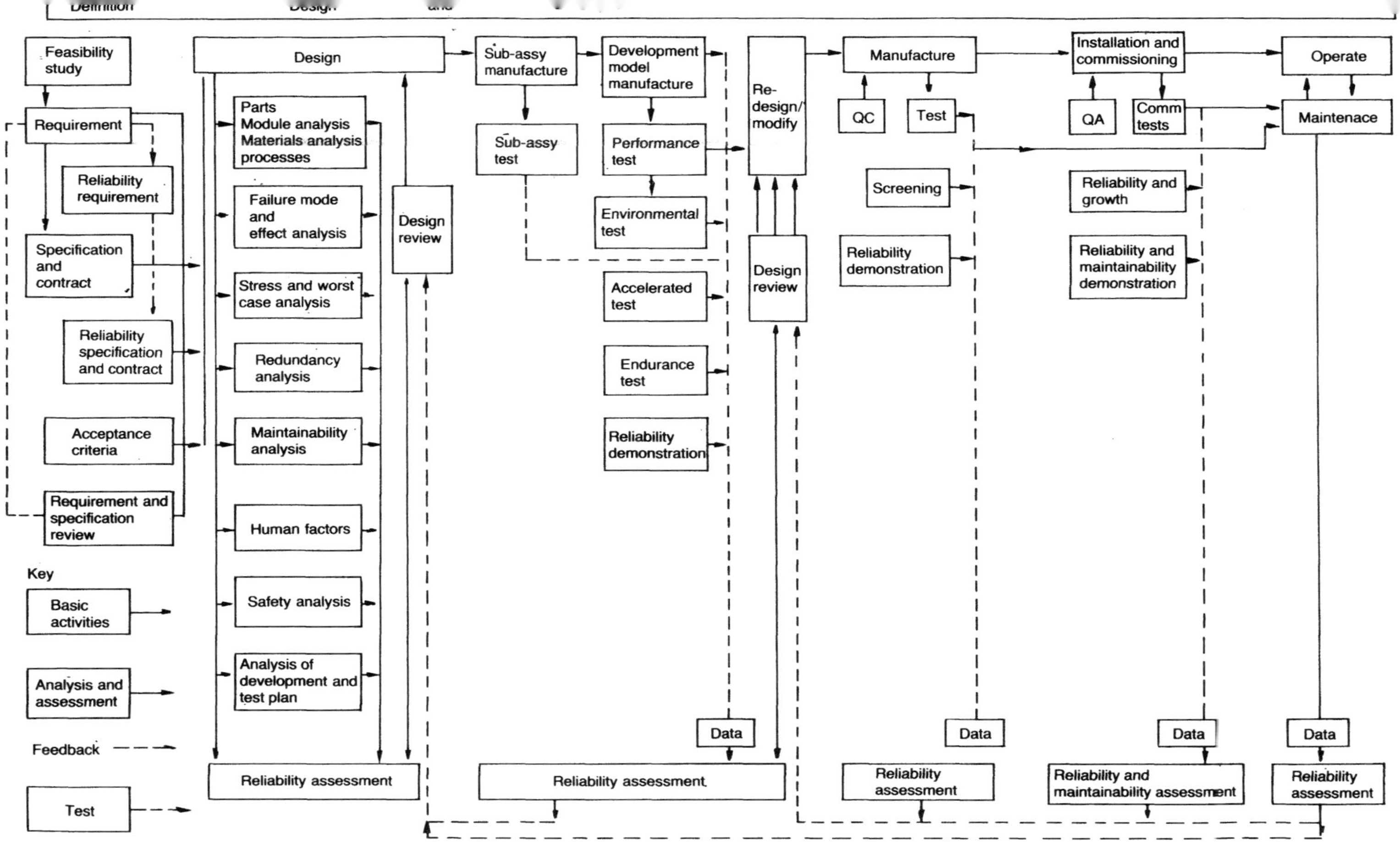

Fig. 13.1 Reliability programme concept (reproduced from BS5760 Part 1 Figure 1 by kind permission of the Director-General of the BSI).

13.1.1 Programme concept

Before making comments on each of these activities, it is necessary first to emphasize the cyclic nature of the whole program, that reliability management is only part of the cycle and that there are many sub-cycles or iterations within each major cycle. What is being attempted is the control of reliability by managerial action. This is analogous to a mechanical or electrical control system in that it requires that feedback signals be compared with set-points. In management the feedback signals are data or estimates of data and the set-points are the specified reliability outcome values. This is illustrated by Fig. 13.1 which is also taken from BS5760 Part 1 where it appears as figure 1. We are grateful to the Director General of the British Standards Institution for permission to reproduce this table and figures.

13.1.2 Evolutionary *v.* revolutionary designs

Designs for products may be evolutionary (developed from the previous, similar, designs) or revolutionary (involving totally new concepts unrelated to previous designs). Evolutionary designs generally have a better chance, other things being equal, of achieving the required reliability in the allotted timescale and within a reasonable budget. However, the introduction of new principles has got to occur sometime, in which case more time and money must be spent on reliability assurance.

The whole design process is cyclic in the case of evolutionary design. In revolutionary design the number of sub-cycles in which successive prototypes and pre-production models are made and tested is usually greater, but sub-cycles appear in both cases.

No design is, in practice, either fully evolutionary or completely revolutionary. There are some elements of innovation or experiment even in the most conservatively evolutionary designs and there is some re-use of old technology and experience even in the most revolutionary new design.

Three examples will be given. In the microcomputer field the step from 386 to 486 technology is mainly evolutionary but the introduction of parallel processing and transputers would be classed as predominantly revolutionary, yet the '486' chip had to be proved and the transputer uses a lot of proven components in its circuits. In the world of railways steam gave way in a revolutionary way to electric and diesel propulsion. Early problems with transmitting diesel power to the rail wheels led to the adoption of electric transmission, the problems of which had meanwhile been solved on purely electrical trains. A further railway example arose in the UK with the development of two new trains. One was named the High Speed Train (HST) and the other the Advanced Passenger Train (APT). The only novelty for British Rail in the HST was the placing of a diesel-electric unit with driver's cab at each end of the train so that engines need not be run round at the end of journeys. The presence of two engines was actually beneficial to reliability because it introduced partial redundancy. The transmission of signals from end to end to control the rear engine proved not to be a problem; something very like it had, after all, been used on purely electric trains for many years. The only other new technology was a modest advance in suspension design to allow a modest (14%) increase in maximum speed from 175 to 200 km/h (110 to 125 mph) over the previous trailer coach design, and a few minor features of the engines. The diesel engine design had been used by the Royal Navy for power generation and minesweeper propulsion for some years but was re-rated at 10% more power for the rail application. The novel features of the HST were extensively and separately tested before final tests of a prototype train were made prior to service. The HST was introduced successfully then on the very easy London to Bristol main line of under 200 km, using eight-coach trains compared with a design train length of up to twelve coaches. As more HSTs became available they were introduced on successively longer routes – London to S. Wales, London to Penzance, London to Leeds and finally London to Edinburgh and on to Aberdeen. The HST has been successful in service apart from some problems with the engines which have been apparent to customers. This may be due to the very intensive use patterns or the 10% power uprating referred to above.

The APT, on the other hand, was a more ambitious project involving a total of 33 new design features and a 40% increase in speed from 175 to 240 km/h over current practice at its inception. The suspension was novel and 'live' in the sense that the coaches were supposed to tilt in bends to counteract centrifugal forces so that higher speeds could be sustained without realigning track. The first prototype was gas-turbine powered. Although British Rail had experimented with gas turbines previously in a small way, this could be counted a revolutionary feature. In the end, however, the train was electric powered because the London–Glasgow main line electrification was completed before it entered service. Although development took a long time it was not very intensive; for years there was only one prototype. With so many novel features it was difficult to conduct tests on one of them whem 32 others were not working very well and likely to stop the run at any moment. The lesson here is to test novel features separately before bringing them together, and not to bring them together until satisfactory reliability was evident in each. Two major problems occurred. First the already late change to electric propulsion had to be redesigned rapidly. The first design followed the HST in having a motor at each end. Unfortunately the rear pantograph did not receive enough current because of the waves set up in the overhead wire by the passage of the first pantograph. The final design had both propulsion motors in the middle of the train with a single pantograph above the driving car and driver's cabs at each end. The second major problem was the continued unreliability of the tilting mechanism. This was really a problem in series reliability; all twelve similar devices had to operate correctly because if even one was inoperable the train had to be slowed in bends. Their reliability quite simply was never good enough to be raised to the twelfth power! Matters were not helped by the decision to introduce the train on the full London to Glasgow run of 640 km in the middle of the worst winter of the decade. This is perhaps the most difficult main line in Britain for bends and gradients. Failures were so frequent that it was soon taken out of service and the project, which was very expensive was abandoned. Considering the good world-wide sales potential of a train capable of 240 km/h over existing track, this was a great pity. An Italian group has simplified the tilt control concept by using the pendulum principle and looks set to produce an adequately reliable train. The press made much of the potential of the APT's tilting mechanism to induce travel sickness. Perhaps British Rail should have been less liberal with the drinks on the ill-fated inaugural run, for there is no history of the test engineers suffering! A lot of aircraft structural technology was transferred to rail by the APT project, most of which was by way of reducing the mass to be moved. This together with the electric propulsion made for good acceleration capability which is an important feature on bendy, hilly routes and especially if stops are included.

It is interesting to contrast the APT's economical lightweight design approach with that of the French *Train à Grande Vitesse* (TGV). SNCF, the French rail undertaking, built brand new separate tracks on major routes to operate the new trains. These new tracks were kept straight as far as possible but gradients were not engineered out as in the earlier Japanese Tokaido lines. Big heavy trains with large traction motors push the TGV to speeds up to 300 km/h on record runs. The pantograph wave problem was solved by running a very fat cable from end to end and having only one pickup point. By building everything very solidly and with reserves of strength and power, high reliability was achieved but at a high price. However, the very high speeds created competition for the airlines and so it was possible to charge almost economic fares for the service which has proved popular with travellers. On journeys of 500 to 1000 km the centre-to-centre times compete with aircraft and trains are more comfortable and reliable. The APT failed commercially because its reliability was not developed to a competitive level, not because there was anything inherently unworkable about the new technology involved in it. It just needed more investment in development and testing.

The same contrast has emerged in comparisons between the US and Soviet approaches to space technology. The US in general uses sophisticated design techniques to reduce the mass of the satellite on the top and the non-fuel mass of their rockets. The USSR in general built things rugged and used bigger rockets to put their heavier satellites into orbit. The relatively low failure rates achieved by both are a tribute to their respective reliability engineering groups.

13.1.3 Competing factors

Reliability (including durability and maintainability where appropriate) is only one aspect of design competing with aesthetic and functional aspects for the limited budget available for the development. Scrap rates, customer complaints and falling sales *do* however indicate that the emphasis given to quality and reliability is not always sufficient. So what criteria are we to use to decide how much to spend on making the design reliable? One might be the life cycle cost approach. With Jenney's (1972) 'equation' in mind,

$$\text{i.e.}\quad \text{Value} = \frac{\text{Quality} \times \text{Delivery}}{\text{Price}},$$

it is reasonable for a given level of performance to express value (for money to the customer) in terms of the total cost to the operator over the whole lifetime, divided by the lifetime, i.e. minimum overall cost-rate. For this approach to be fair *all* the costs must be included. By taking account of the need to buy extra items or to forgo profits during breakdowns, we can reduce reliability and maintainability to monetary units. In fact that life cycle costs include elements for design, manufacture, preventive maintenance, repairs, loss of utility and a disposal cost which may be positive or negative.

The life cycle cost approach, sometimes called terotechnology, allows tradeoffs to be made between different aspects of Design on a common base of money. Possible tradeoffs might be between any of the following factors:

> Reliability, Durability, Preventive Maintenance, Quality, Maintainability, Price, Ease of Manufacture, Early Delivery, Maintenance Costs Saleability and Performance.

One has only to think of the factors which affect one's own personal purchase of a car, or other consumer durable, to see how this process might be formalized by balancing costs and benefits. In fact the principle is really the same as that used in a socio-economic cost-benefit study. Figure 13.2 illustrates in a general way where tradeoffs might be made in a project. Every project is subject to different emphases and constraints however, and its tradeoffs must be thought out as a separate exercise. The general idea though is to give the designer a multidimensional envelope of possible latitude in the design.

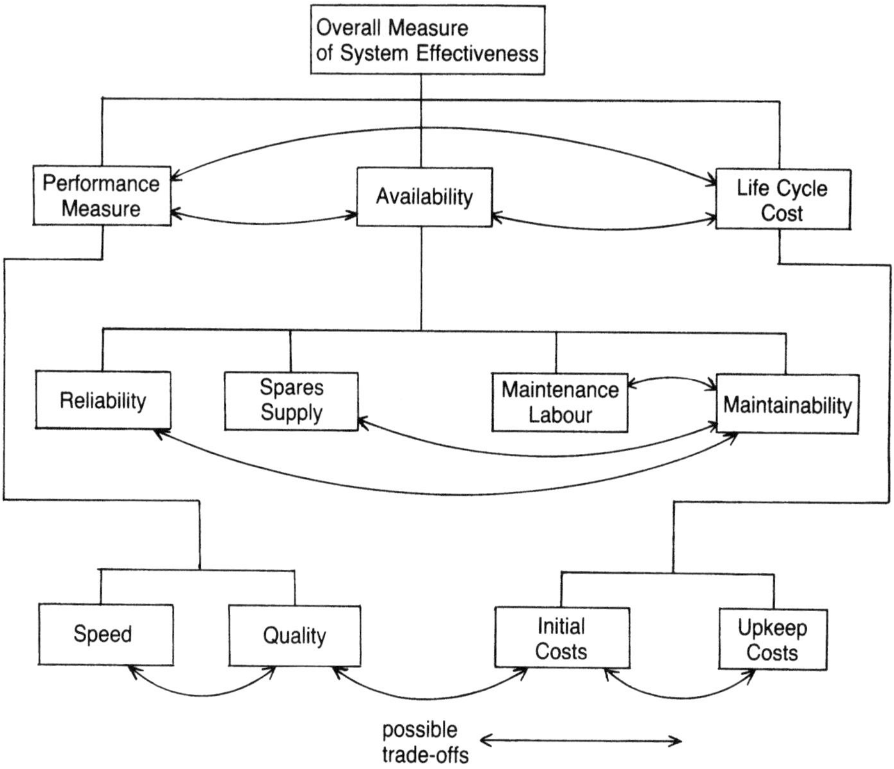

Fig. 13.2 Illustration of the trade-off principle showing some possible trade-offs.

13.2 RELIABILITY PROGRAMME

In this section the four project phases of a reliability programme, *definition*, *design and development*, *production* and *function and maintenance*, will be discussed to more detail. However, it should be emphasized that these phases do not stand alone but are part of cycles within the reliability programme and within the overal project.

13.2.1 Definition phase

At the start of a project it is necessary to lay down reliability and maintainability (R&M) objectives which are sensible and feasible. Very often feasibility studies include study of R&M possibilities and lead to a statement of R&M requirements or a formal contractual document setting numerate targets. At this point an R&M program should be drawn up integrating R&M activities with the rest of the project management development plan.

Definition may be carried out by the customer, the contractor/producer or by both in conjunction. In defence work the customer usually proposes what the new weapon or system should be able to do. The statement of requirements, as the proposal is usually called at this stage, contains minimum standards for both performance and reliability related factors. The basis of the proposed design, the technology, will be stated usually, and may have come from a defence R&D establishment. However, turning the proven principle of operation into a working system to be manufactured on a production line is usually the business of the contractor.

Where the customer is not involved in the specification as in the case of manufactured goods to be offered in competition with other similar goods in an open market (e.g. motor cars or consumer durables), it is essential to success that a group within the company acts on behalf of the customers and their interests. The design and production of such items is inevitably a compromise between quality and ease of manufacture. In order to capture a viable share of the market it is necessary to have an acceptably reliable product. This will not occur without the considerable effort and expense of a reliability program which starts at this first phase of the cycle.

We take the items in Fig. 13.1 under 'Definition Phase' in turn.

The reliability feasibility study is part of the broader feasibility study undertaken to examine the overall technical prospects of the project. The basic function of the study is to determine first what would constitute acceptable reliability for the product and under what conditions it should be measured, and then to decide whether this is possible within the likely development budget. For a complex item it is usual to draw a reliability block diagram (RBD) at this stage and to fill in numbers for as much as can be predicted confidently, in order to assess what can be allowed for the less predictable features of the design.

The statement of reliability objectives and requirements emerges from the feasibility study as a compromise between what is felt to be wanted by the customers (or their advocates within the company) and what is considered to be feasible technically within the budget. This may really be a working document for a preliminary contract, formal or informal to work out a reliability specification and contract clauses related to reliability.

The reliability specification and contract clauses related to reliability may be a legal contractual document formalizing agreements with the customers or it may be an internal and informal document setting limiting values and envelopes of acceptability for the various reliability-related factors. It is most important that the specifications include consideration of the following features:

1. the reliability block diagram or other representation of redundancy and dependence;
2. the levels achieved by competitors;
3. the state of development of components (affects uncertainty factors);
4. intensity and duration of duty;

5. acceptable level of preventive maintenance;
6. limiting environmental conditions;
7. constraints on performance, size, weight, etc.;
8. interaction between safety and reliability.

Section three of BS5760 Part 1 gives detailed advice on the writing of specifications for reliability and related contract clauses. Picking the essential and notable features from this text we draw the reader's attention to the following:

1. failures must be defined precisely and in as many contexts as are relevant;
2. a decision has to be made as to whether to specify only the results and how they will be assessed or tested, or alternatively to specify procedures for each stage of product development which aim to assure the specified results;
3. quantitative specifications must always include consideration of statistical confidence intervals;
4. specific test conditions and reliability under test may differ from service expectations.

13.2.2 Design and development phase

In Table 13.1 is a long list of R&M activities to be carried out during design and development and before marketing the product. Some of these will be studied later in more depth. The main points in the *management* of this stage of the process are as follows.

1. Have a program and stick to it.
2. Build in R&M from the start (it is much more difficult to put it in afterwards).
3. Trade off by all means but ensure that any lowering of quality standards is truly compensated elsewhere. Avoid late alterations.
4. Take the customer's part in internal discussions. The reliability section's job is to get him value.
5. Involve customers if appropriate and otherwise study their complaints and the spares they have used.
6. Ensure that numerate test criteria are decided before the tests take place and are met or deliberately altered by executive decision in full knowledge of possible consequences.
7. Pay particular attention to novel features whether of material, manufacturing process, operation, maintenance or design principle.
8. Don't confuse safety with reliability but make sure that both are adequately covered.

The principal techniques used in this phase to improve the product's reliability are design review and failure modes, effects and criticality analysis (FMECA). The cyclic nature of development has already been mentioned.

Design reviews take place at meetings between the designers and others. The others should include the reliability section, some more experienced designers not engaged in this project, customer's representatives if appropriate, and production and quality personnel. The object of the meetings is to review progress in getting the design within specification and budget, or as good as it can be in all respects, including reliability. The meeting should be structured so that actions with deadlines are placed upon participants and action items from previous meetings are reviewed. It is customary in some circles for these meetings to be chaired by the reliability engineer, first because that position is seen as neutral between customers and designers, and also between quality and production, and second because the reliability section has most to lose if the program gets behind schedule or ahead of budget. To explain the second case: it is the reliability program which usually gets cut if time or money must be saved.

Meticulous development of reliability can make the difference between success or failure of a product in the market place or at a higher level between selling thousands and selling millions. An example of this is afforded by the cases of the Volkswagen 'Beetle' and the Austin/Morris Mini. Both cars were brilliantly conceived and ahead of their time but it was its high reliability and durability that kept the Beetle in production so long (over 40 years). It outsold the Mini by an order of magnitude. Moreover, the Mini never made more than a marginal profit for its builders and ran at a loss for several years. Phil Crosby (1979) says that 'Quality is free'. Reliability and its related factors are all elements of quality to which this adage applies with even greater force. This is because their values are largely determined at the design and development phase when it costs little to make changes that would be very expensive later on in the product cycle.

The two notional graphs at Figs. 13.3 and 13.4 illustrate this point.

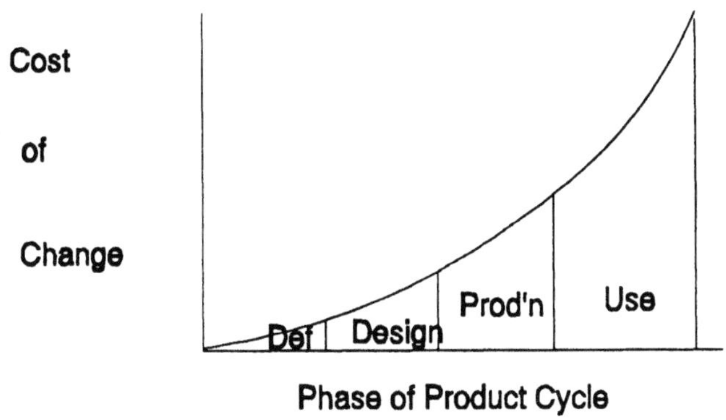

Fig. 13.3 The cost of making changes to a product grows rapidly with the phase of the product cycle.

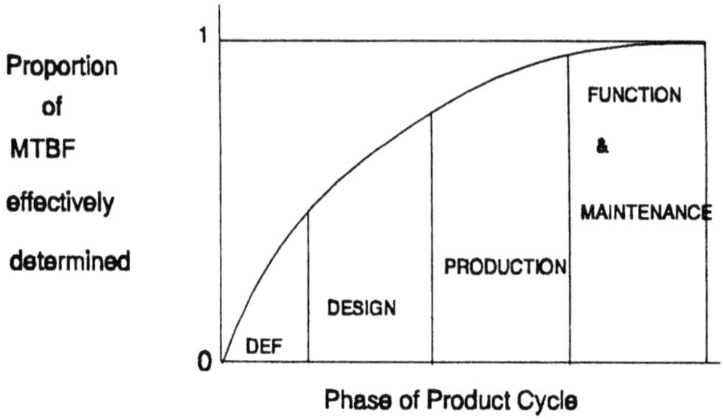

Fig. 13.4 The efficiency of reliability improvement efforts is greatest in the early phases of the product cycle.

Because it becomes more expensive to make changes as the project advances through the phases of the product cycle it implies that the final system reliability can be adjusted less and less and is mainly and effectively determined in the early stages. The biggest headache is that nobody *really* knows what it is going to be in the end.

(a) Failure modes, effects and criticality analysis (FMECA)

However, failure modes analysis can help us to improve the final result by anticipating individual modes of failure. Those that are considered serious threats to the success of the project and also those which are easily and cheaply dealt with can then be eliminated or severely reduced by design changes. FMECA is basically an organized and tabulated statement of the details of the latest assessment of the failure modes likely to occur with the new design and their effect on operation. It is useful to add notes about what it is proposed to do to reduce the failure incidence, and if it is FMECA rather than FMEA then an estimate of frequency of failures and of mean time to repair or of the distribution parameters for times between failures and times to repair will be included. Another feature of criticality is a subjective scale of seriousness of particular failure modes on the basis of a combination of features such as safety, repair cost, operational effect, and legal considerations.

FMECA is continuous during design and becomes more detailed and more accurate as the design progresses. The amount of detail increases with each iteration of the design phase cycle. The accuracy is improved as more detail is added and more tests are completed. It is therefore a good idea to keep the master FMECA documentation in pencil on large sheets of paper (at least A3) or on a computer spreadsheet so that detail is not omitted for want of space. Copies of successive documents form a record of reliability achievement and decisions which will prove valuable later in analysing service failure data and resisting cheapening design changes likely to cause reduction of reliability.

An FMECA ideally is built upwards from component data by considering first individual modes then modes due to 2 or 3 components' interactions and so on. This *bottom-up* method is effective but tedious and expensive and has now been all but abandoned by even the US military.

A *top-down* FMECA starts at the other end with the effects and considers product failure as the sum of assembly failures. This means that the analysis can be stopped at a level where satisfaction of specification for an assembly has been assessed. This method is not so thorough but is obviously less wasteful of time and effort which may be in short supply. Provided that stress levels are the same or less, it is reasonable to assume that a sub-assembly will behave much as it did in a previous model rendering component analysis unnecessary. However, if stress levels have been raised it is difficult to extrapolate failure frequency without tests or deeper analysis. Local bottom-up analyses of particular assemblies in a generally top-down analysis are a good compromise.

Column headings on the pro-forma will be subject to variation from industry to industry but may include the following:

1. assembly;
2. sub-assembly (sub-sub-assembly);
3. parts/components involved;
4. description of failure mode;
5. symptoms (particularly warning events);
6. primary effect on operation of:
   (a) sub-assembly,
   (b) assembly,
   (c) complete product;
7. classification on severity scale, for example:
   (a) Class 1 – loss of life, loss of product, legal liability,
   (b) Class 2 – extensive loss of use,
   (c) Class 3 – minor loss of use or reduced quality of service,
   (d) Class 4 – any less serious failure;
8. failure rate/distribution parameters;
9. how estimated, e.g. previous model, tests, service data;
10. repair method;
11. repair time distribution parameters/MTTR;
12. proposed action:
    (a) to reduce incidence,
    (b) to reduce down-time,
    (c) to reduce cost,
    (d) to give better warning;

13. secondary effects on operation of assembly and product (with classification rate, repair time, costs, etc.);
14. relevant PM routines;
15. remarks – to include latest progress report and any deadlines, etc., imposed by design review meetings;
16. date entry last revised.

The FMECA documentation and design development program together provide effective control of reliability development for a new or revised design and a useful record of the way the final design was formulated. As such it is important to keep records of both aspects for future reference in planning new products.

The assessment of frequency of occurrence of failure modes may be absolute or relative, depending upon what data are available in respect of components. The purpose of criticality analysis is to highlight those failure modes which must be eliminated or reduced in frequency. Criticality is a combination of severity of effect of the mode and its (relative) frequency of occurrence. It is usual (and in US military contracts obligatory, see MIL-STD 1629 (1989)) to make a grid as shown in Fig. 13.5 and to write in the identification codes of all failures in the appropriate squares. Criticality is then measured roughly by how far along the diagonal the failure mode appears. Thus a mode in square E1 would be regarded as about equally critical as one in A4, likewise B3 and D1, or A2 and B1, or C4 and E2. The frequency scale may alternatively be in terms of order of magnitude of the ROCOF. The Military Standard permits the definitions in the severity code to be varied to suit different applications provided that these definitions are carefully stated.

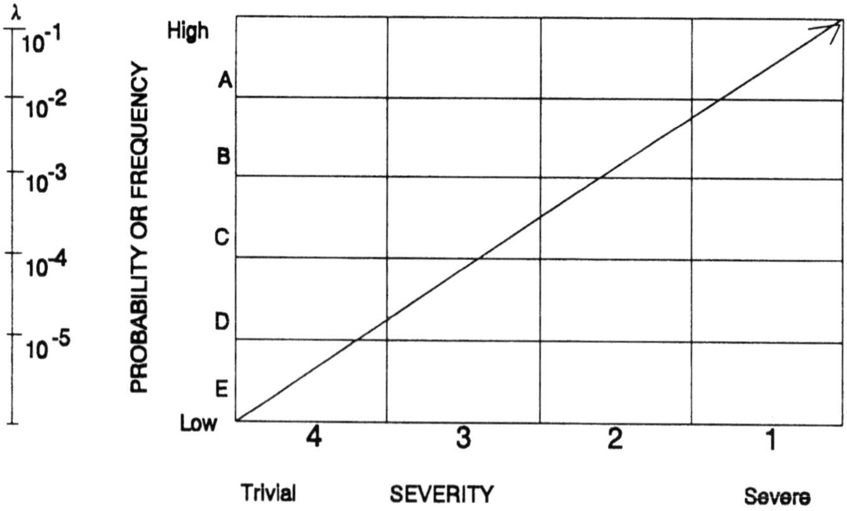

Fig. 13.5 Criticality chart.

		ITEM(S) REF DRAWING No MISSION DESCRIPTION DATE SHEET OF FAILURE MODE & EFFECT ANALYSIS			
FAILURE MODE I/D No	FAILURE MODE DESCRIPTION INCLUDING (POSSIBLE) CAUSES	ITEMS INVOLVED I/D No	DESCRIPTION OF FAILURE EFFECT (a) LOCAL (b) NEXT LEVEL (c) COMPLEX	HOW FAILURE IS DETECTED INC WARNINGS	PROPOSED DESIGN, MAINTENANCE &/or OPERATIONAL INSTRUCTION CHANGES TO PREVENT/REDUCE INITIAL FAILURE AND EFFECTS

Fig. 13.6 FMEA proforma.

MIL-STD 1629 gives a method of calculating a criticality number which is accommodated in the pro-forma example at Fig. 13.6. This form is designed to be placed to the right of the corresponding FMEA form at Fig. 13.5. Taking the columns in turn, first the failure-mode identification is repeated (in case the pages get out of order!). The next column is for special factors. This might be used to state whether repairs were possible without interrupting service, or at sea in the case of ship-borne equipment or for some other important but special factor affecting the criticality of a failure mode in the particular application. The next two columns are for the four-point severity and five-point probability of frequency codes. The next four are for the calculation of criticality number in accordance with MIL-STD 1629. This is done as follows:

1. β is the conditional probability of loss of mission given that the failure mode has occurred.
2. $\propto$ is the proportion of the total ROCOF of the component covered by the I/D No. which relates to the failure mode under consideration.
3. λ_p is the total ROCOF of the component covered by the I/D No. or the latest assessment thereof.
4. t is the total time the component is operational during the mission. This may be the whole mission period or only some part or parts of it.
5. $C_m = \beta \propto \lambda_p t$ is the criticality number.

The use of the word 'mission' betrays the military origin of this calculation. C_m is actually the expected number of failures per mission in that mode which result in mission loss. Because they are expected numbers of failures rather than probabilities it is possible to add the mode criticality numbers together to obtain item, assembly and even system criticality numbers. If the total criticality number is C_r then the probability of mission failure is given by

$$P_{mf} = 1 - \exp(-C_r). \tag{13.1}$$

The value of β in each component may be 0, 1 or possibly some fraction in between. Fractional values arise when system failure or mission abandonment will only occur if some other conditions exist. These other conditions may relate to environmental factors or mode of operation or human failure or failure elsewhere in the system.

There is much to be said for making the 'level of indenture', as the MIL-STD describes the level at which analysis is made, the same as the level at which data are available. In that case, C_r for the item will be correct so that, even if some of the intermediate C_ms for modes are partly guesswork their sum is known and they can be made to add to the correct figure.

Criticality number is actually a misnomer; as stated above, it is actually an expected number of failures in a fixed time. There is no reference to the severity element of criticality except insofar as that is contained in the definition of the factor β. However β merely restricts the definition of failure to mission or system failure: it does not examine the safety or even the downtime aspects of these failures. It reflects the preoccupations of the military with short-term (mission) reliability and mission success rather than availability. In civilian applications, the reliability engineer should feel free to define and use other measures of criticality more appropriate to the circumstances.

(b) Analysis of parts, materials and processes

The production people present at design review meetings will inevitably be interested in how each component is to be made. There should be, ideally, a production engineer on the design team during this phase to advise on these matters. For costing purposes, an analysis of which parts are to be made and which bought in, and of the materials and processes to be followed is very necessary. Until it is drafted it is not possible to carry out important quality and reliability actions.

A second FMECA should be conducted in which production processes to be used by the proprietary or contracting company and any sub-contractors are analysed to assess what could go wrong at each stage of production and to advise on what quality control (QC) measures would be necessary to minimize the number of defectives passed on down the line. It is vital that QC is applied where necessary and only where necessary, since it is a non-productive activity whose expense must be justified. QC is not always statistical QC; quality can be controlled by careful delineation of the process parameters, by buying more capable

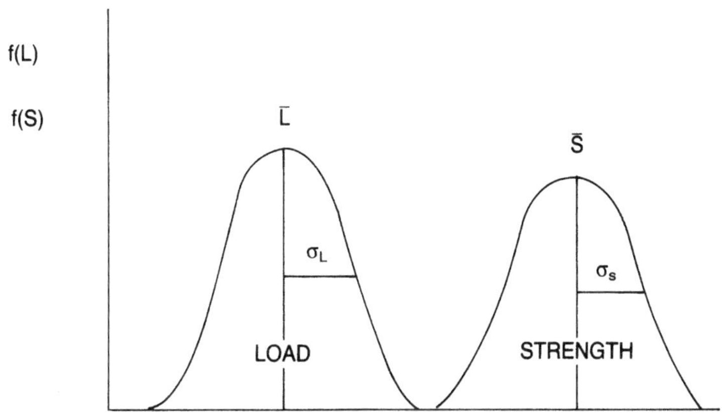

Fig. 13.7 Load and strength variability.

machinery, by redesigning the components, and by regular maintenance of the machinery. Such *preventive* measures are usually preferable to SQC and other *appraisal* methods because the long-term costs are generally lower.

The production FMECA will also reveal cases where current machinery will not meet quality standards and investment must therefore be made in new and better machinery. Most investment will have been anticipated before this but the production FMECA will confirm and modify the schedule of new equipment to be obtained.

Fault trees may be used during this phase to elucidate some of the trickier dependencies where a *reliability block diagram* (RBD) is more difficult to draw or inadequate. Fault trees have the advantage that computer programs exist for their construction and solution. Success trees may also be used (Chapter 9).

The safety program also usually relies upon fault trees for the analysis of ways in which dangerous situations may arise and their probability of occurrence. It is important to consider safety and reliability requirements *together* as safety considerations will affect the apportionment of the maximum unreliability and unavailability permitted in the specification to the various parts and assemblies making up the complete system. This is because although most accidents involve either human error or 'acts of God' as well, there is usually also an equipment failure which in many cases can be regarded as the key event. Many forms of accident are much less likely if the equipment is operating correctly.

(c) Stress and worst case analysis

As this name implies, stress and worst-case analysis involves investigation of extremes of stress and operating conditions and the effects they may have on component and system reliability. It is usual to take cases somewhat beyond the limits of the range expected in order to be sure. It is vital that where any doubt exists, the minimum strength, in the most general sense, should be related to the maximum corresponding loading (also in the most general sense). In some fields of work, e.g. pressure vessels, the maximum loadings may be determined by Codes of Practice. As stated previously, it is most important to check that nothing for which reliability figures are being assumed from previous use is being stressed more highly in the new design than in the old. Beware, for example, of the salesman who once approached a production engineer with details of a 'new' faster lathe which turned out to be the old one with a bigger motor. The reserves of strength to cover the increased stress may or may not be present and the margin has certainly been reduced.

Carter (1972, 1988) has suggested that the traditional factor of safety (FOS) be replaced by a margin of safety (MOS), reflecting the part played by the variability of the load and strength as well as their average

ITEM(S)
REF DRAWING No
MISSION DESCRIPTION
DATE
SHEET OF

CRITICALITY ANALYSIS

FAILURE MODE I/D No	SUPPLY INTERRUPT	SEVERITY CODE	PROBABILITY CODE	CONDITIONAL EFFECT PROBABILITY β	FAILURE MODE RATIO α	FAILURE RATE ESTIMATE $\lambda\rho$	OPERATIONAL TIME PER MISSION(t)	CRITICALITY No. Cm= $\beta\alpha\lambda\rho t$	REMARKS	DATE OF LAST AMENDMENT OR ENTRY

Fig. 13.8 Criticality analysis proforma for use with Fig. 13.5.

values. Failure occurs when $S \leqslant L$. The distribution of S-L has mean $\bar{S}$-$\bar{L}$ and s.d. $(\sigma_S^2 + \sigma_L^2)$. The MOS is defined as the reciprocal of the coefficient of variation of this distribution of differences

$$\text{MOS} \equiv (\bar{S} - \bar{L})/(\sigma_S^2 + \sigma_L^2). \tag{13.2}$$

Note that if $\bar{S}$ and $\bar{L}$ are both increased by the same amount, MOS is unaltered but FOS is making it a poor measure of reliability.

A further dimensionless comparator was also proposed by Carter, namely a measure of relative roughness of loading γ

$$\gamma = \sigma_L/(\sigma_S^2 + \sigma_L^2). \tag{13.3}$$

Quality control can reduce σ_S but σ_L is determined by the use to which the component is put. Both may increase with age and $\bar{S}$ will fall with age in most failure mechanisms. By simulating these changes according to various models a robust general shape is obtained for the graph of hazard rate versus the declining MOS.

It turns out that as MOS falls from a high value, the incidence of failure increases very rapidly once a 'period of grace' has passed and the L and S distributions overlap. MOS will fall with age as the component wears or fatigues or corrodes away. Unreliable components are those with insufficient initial margin or which suffer very rapid deterioration in MOS. Most mechanical components outlast the system in which they are fitted, only a few exhibit low reliability. The 'worst case' occurs as some chosen lower limit of S say $(\bar{S} - 3\sigma_S)$ walks into the maximum load say $\bar{L} + 3\sigma_L$ with time. If they overlap at $t = 0$ then the component has no intrinsic reliability, no period of grace, and a proportion will fail quickly in service.

This analysis justifies both stress and worse case analysis and the careful reliability testing of new component designs. We have only to be fractionally in error in our estimates of $\bar{S}$, $\bar{L}$, σ_S and σ_L to have an unreliable component. Luckily the sensitivity also means that usually only a little 'beefing up' is required to put matters right.

Certain components though must, for safety or cost reasons, be designed to break before others, e.g. the half shafts in a car will fail before a wheel comes off or the differential breaks a tooth. In such cases, stress and worst-case analysis is even more important for all the components of the sub-system which might possibly fail.

All the above remarks apply more especially to mechanical components where γ is high than to electronics where γ is low.

(d) Maintainability program

In products where availability rather than reliability is of paramount importance, it will be necessary to look closely at the maintainability. The examination of competing factors (see above), and the subsequent trading off during the definition phase of the project is the start of this procedure. At that time, targets for maintainability or availability will have been set. The analysis of redundancy will also affect the issue. Maintainability also affects the *reliability* of redundant systems which can be a very important point.

It is essential that components which will have to be renewed because their unreliability cannot be improved should be accessible to the maintainer so that availability is sustained. In electrical and electronic work they would plug in rather than being soldered and consideration should be given to means of identifying the faulty component or board as well as of exchanging it quickly. Diagnosis time is otherwise commonly greater than actual maintenance time. Similarly in the mechanical field, it should not be necessary to disassemble the whole equipment to renew frequently failing components and those subject to PM.

Another common failing is not to allow sufficient room around a component to permit its renewal or servicing by the best possible methods. For example, heat exchangers which need frequent cleaning or repair must be located such that the tube nests can be drawn right out without disturbing the shell. In ships and aircraft it is no good to schedule repair or maintenance by replacement of the whole equipment if the equipment will not pass through the relevant doors and hatches. Special trials may have to be mounted to find out what the minimum distances required actually are so that detailed layout can be settled.

(e) Human factors

Operators and maintainers, being human, require special consideration in achieving a reliable design. Because their physiology and psychology varies yet cannot be changed, the machine or system must conform to their needs if reliability is to be maximized. Operating positions and control rooms therefore require careful ergonomic design to ensure that operators are presented promptly with the required information through instruments or their own senses to allow them to respond in a correct and timely fashion to emergencies as well as routine occurrences. The human factor has the potential to improve reliability, but only if this is so. It also has the potential to worsen reliability if, for example, operators are uncomfortable or maintainers cannot see what they are doing. This is because, being human, they become frustrated and react with responses that vary from carelessness to downright vandalism depending upon how badly the ergonomic work has been done.

The other major human factor is training. It is necessary to consider how much training will be required for both operators and maintainers. Recall the bathtub curve and the effect on the ROCOF of the training of operators and maintainers. The bathtub curve theory shows this effect petering out after an initial period, but if nothing is done, the later constant value of the ROCOF will usually be higher than needful because of the frustration factor referred to above. When CNC machine tools were introduced in the UK it brought to a head a phenomenon that had been building up for a decade or more among the electricians. Some of these tradespeople were expected to service the new electronics with no training beyond that of a line electrician. The situation became so bad that, in default of the employers doing anything, the Electrical Trades Union found itself running electronics and control courses for them. Companies simply did not realized the need, and so the availability of plant suffered severely. It was a very serious matter; it broke some manufacturing companies altogether, because they could not get the expected return on their considerable investment in CNC due to delays and foul-ups in maintenance caused by simple ignorance. Another likely effect of lack of training is more frustration, which may then lead on to carelessness or vandalism, which is often rationalized by the workforce into opposition to the new technology.

(f) Safety

The third aspect affecting and affected by the person–machine interface is safety. The importance of the human links in the causative chains which should be identified in the safety program cannot be overemphasized. Two techniques are considered useful, namely hazard and operability studies and fault-tree analysis (FTA) (Chapter 9). The hazard and

Table 13.2 Checklist of keywords for safety studies

Operators	*Inputs*	*Environment*	*Failure modes*	*Solutions*
What	Material	Pollution	Wear	Strengthen
Where	Process	Waste disposal	Corrosive	Duplicate
When	Temperature	Noise	Fatigue	Control
How	Flow rate	Weather	Creep	Enclose
Who	Pressure	Lightning	Chemical	Isolate
Too much	Maintenance	Toxic	Burst	Dispose
Too little	Speed	Atmosphere	Collapse	Reduce
More of	Power supply	Riverine	Implode	Increase
Less of	Cooling	Ground water	Explode	Redesign
None of	Chemistry	Acid rain	Detonate	Fail safe
Part of	Physics	Vapour cloud	Fire	
As well as	Start-up		Output loss	
Other than	Shut-down		Poor quality	
Reverse	Inventory			
	Quality			
	Reliability			
	Availability			

operability study (HAZOP) technique originated in the chemical industry but there is no reason why it should not be applied to other manufacturing systems and even products. It is a more broad-brush procedure than FTA. In fact, it should precede FTA because it is essentially a quick way to identify the top events for a safety FTA.

One company uses a sheet of keywords as in Table 13.2 and the schematic drawing of the system at the first few meetings of a committee of experienced designers, operators and safety and quality control personnel. Using Edward de Bono's technique of brainstorming, assisted by the list of keywords they go round the schematic from raw material input to product output and then round the table thinking aloud all the things that could go wrong and affect safety, however extraordinary or even far-fetched. In this procedure extroverts are needed, for it is no good becoming embarrassed over suggesting impossible scenarios that are destroyed by colleagues. That is the whole point – to explore all the possibilities in order to pick out those that really might cause trouble, which must be then subjected to detailed analysis, usually by FTA.

BS5760 Part 1 places the safety program in the design and development phase, but there is much to be said for starting the process during project definition. In many countries and for some projects, safety or environmental impact studies are required to obtain planning permission.

The same company envisages a total of six stages to the Safety Program for new chemical plants:

1. brainstorming as above at the feasibility study stage;
2. preliminary FTA to be completed for planning permission and internal financial sanction;
3. frequent meetings of the committee to identify hazards which are then passed to experts for solution/advice;
4. inspection during construction to ensure it is being done properly;
5. repeat 4 before startup to ensure cleaning up, guard rails, notices and other routine safety matters are done;
6. review after about 1–2 years service.

Note: we had not intended to say so much about safety, but it is necessary to put the powerful technique of FTA into perspective regarding one of its principal uses. We claim to have done no more than that; this book is *not* about safety except insofar as it bears upon reliability.

(g) Redundancy analysis

Redundancy costs money. It is often vital that it is used to best effect to raise reliability. The simple rule is that it will have the most effect in the least reliable stages. However, if cost and/or mass are also considered and limits exist on either or both, working out the best stages to duplicate can be more complicated.

$$1\bigg/\left[1+\sum_{i=1}^{N}\frac{(1-A_i)}{A_i}\right]=A_s$$

is to be maximized subject to

$$\sum_{i=1}^{N} m_i j_i \leqslant M,$$

where $m_i j_i$ is the mass of j_i extra items in stage i and M is the total extra mass permissible and/or

$$\sum_{i=1}^{N} C_i j_i \leqslant C, \tag{13.4}$$

where $C_i j_i$ is the cost of j_i extra items in stage i and C is the total extra cost available

$A_{i,j} = a_i$ for $j = 0$ extra items in stage i,

$A_{i,j} = 2a_i/(1 + a_i^2)$ for $j = 1$ extra item in standby, 2 repairmen,

$A_{i,j} = 1 - (1 - a_i)^k$ for $j = k - 1$ extra items in parallel
$k \leqslant 1$, k repairmen,

etc. as in Chapters 2 and 3.

Note that the suitability of standby or parallel redundancy will be a technical question to be decided before proceeding to calculations.

This is a multi-stage decision process. First we write down all the technically possible single additions of redundant items which lie within the constraints C and M. It is axiomatic that we should start by considering for inclusion the single acts which give the best return in terms of increased system availability for the least cost and mass. This can be done by ranking the individual additions in descending order of

$$\Delta Q_i = \Delta A_{si}/\text{Max}[c_i/C, m_i/M], \tag{13.5}$$

where ΔA_{si} is the increment in A_s if that individual addition were the only one made and it was made in stage i. Note that the maximum increment in A_s occurs always at the least available stage and that this remains so for the new least available stage after each allocation of a redundant item has been made. However, the most effective solution may in theory have to omit even the addition of an extra item at the least available stage if that addition is very costly and/or massive. *For ranking purposes* it will be sufficient to approximate that the availability of the stage under consideration multiplied by that of the rest of the system is roughly equal to A_s. Then it is easy to see that

$$\Delta A_{si} \approx A_i\,\Delta A_i/A_s, \tag{13.6}$$

where $\Delta A_i = A_{i,1} - A_{i,o}$ is the increment in stage availability, following the first redundant item.

Besides the ranking in terms of ΔQ, a ranking in descending order of ΔA_{si} should also be obtained, and in complex cases extended to include higher values of j, the number of extra items, for some of the least available stages. Where the two rankings coincide or nearly so there is no real problem in finding a good feasible solution within the constraints. However, this will only be the case where the least available items are also the cheapest and/or lightest. Each ranking list should be pursued downwards until a constraint is encountered (cost or mass). Some items will appear on both these priority lists. These become part of the solution. If none appears on both lists (most unlikely) then take the item giving the best increase in availability within the constraints C and M. The problem is then reduced to finding the best solution for reduced C and M over a reduced list of possibilities. The above procedure is then iterated with the possibility of a further item being considered also for those stages which had redundancy allocated in the previous round. The suggested procedure is based upon no redundancy being allocated before the first round; that is we start always with a basic design with a series RBD. In each round there will inevitably be a possibility of 'filling up' with a light or low-cost item to use the last of the remaining cost and/or mass. There is actually little point in doing this on every round: it should be left until, at last, there is only one possibility (or none) fitting the remaining C&M values. At this point an executive decision must be made to either exceed a constraint or not to utilize it fully.

N.B. The above procedure may not always produce the best overall solution, but it was better in trials of simple cases than allocation on availability alone, in the presence of other constraints. Research now in progress may result in a more satisfactory true optimization method for this situation.

In practice, engineering or safety considerations may override the computed solutions. Also in some applications, maximum productiveness, maximum reliability over a stated time or minimum ROCOF may be the criterion of excellence rather than availability. It is of course vital to be clear what criterion is to be used at the outset. This will have been a part of the definition phase of the project.

In many cases failure is self-announcing but where this is not so BS5760 Part 1 enjoins as part of redundancy analysis that the provision of indicators that failure has occurred to a redundant unit be considered, since only then can the full potential system availability be realized. As the standard says, this is essential on items with high criticality ratings due to safety considerations, etc. It is also relevant in some cases to question the assumption that parallel operation does not affect the ROCOFs of the units in question; e.g. reduction of load may reduce electronic component hazard rates but actually increase the ROCOF's of centrifugal pumps.

(h) Testing program

The need to plan for adequate time for testing has already been mentioned, as has the need to test novel features thoroughly and separately. A lot of the necessary testing of new product designs is to develop performance. However, all reliability data arising from all tests should be kept for possible later use even when the trials are not specifically to test reliability. The following analysis of the types of test that may be required and how to conduct some of these should be read in conjunction with the sections on testing in Chapter 11. Figure 13.9 shows several separate ways in which tests may be classified.

Thus we may have, for example, an accelerated endurance test of a prototype to assess its reliability, or a field test of a production model to demonstrate its endurance. All failures and other incidents occurring in development testing should be noted and consideration should be given to deliberately amalgamating performance and reliability development testing into one programme with twin objectives, using TAAF and the Duane approach, detailed in Chapter 11, to document reliability growth, followed by a Wald sequential test to demonstrate that the required MTBF has been achieved with the required level of confidence.

During product development, accelerated, field and environmental tests are often conducted with the objective of finding the weak points in

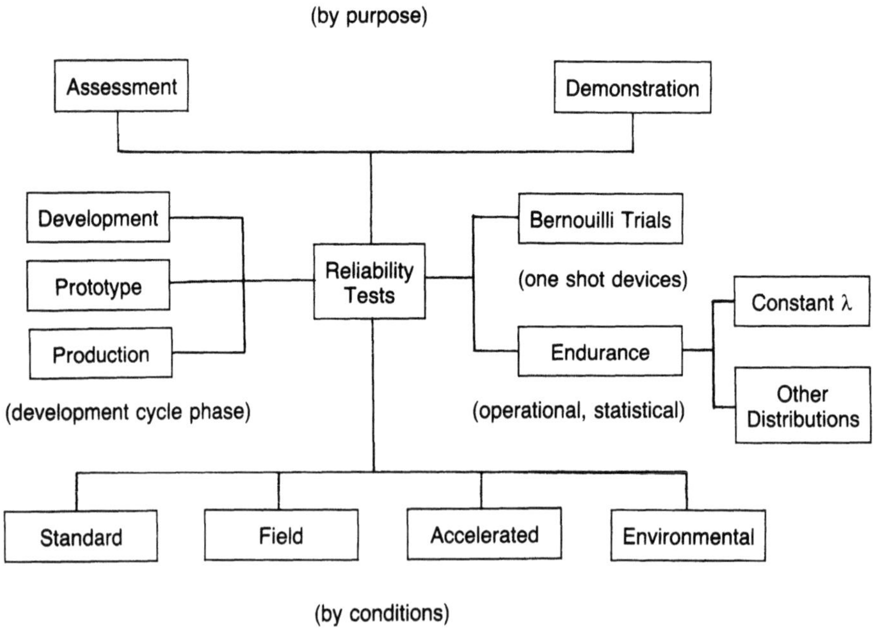

Fig. 13.9 Classification of reliability tests (adapted from Fig. 18 of BS5760 Part 2 which was supplied to the BSI by DJS).

a design rather than assessing current reliability. Indeed, it is axiomatic that reliability development testing should involve some failures if it is to be useful to the designers, even if this means quite severe overstresses being applied in order to make something fail in a reasonable time. It is important to know which will fail first out of a number of components because this will affect safety (possibly) and maintenance (certainly).

When it is not known how much stress (in the most general sense) will cause failures, accelerated testing may be conducted according to the following general method. First choose an environmental or other factor which can be progressively enhanced and which is expected to cause a fall in the mean time to component failure without *changing the mode of failure*. Above a certain limit, it may be possible to estimate that failures will occur by a different mechanism. For example, electronic components are commonly assumed to follow the Arrhenius law, which means that the hazard rates increase exponentially with absolute temperature. However, above a certain temperature, melting of the solder or disintegration of the base material of the circuit board will become dominant and the relationship becomes invalid. Below the limit we can accelerate the failure process by successively stepping up the stress until failures occur.

This is more efficient in the use of test specimens than the more usual method of testing at a number of different constant stress levels. It will be clear from Fig. 13.10 that the total time to complete a test series by stepping up a radius from the origin is going to be less than that to complete a similar series at constant stress. The analysis can be made by making Weibull plots of $(s^2 + t^2)$ against mean or median rank of

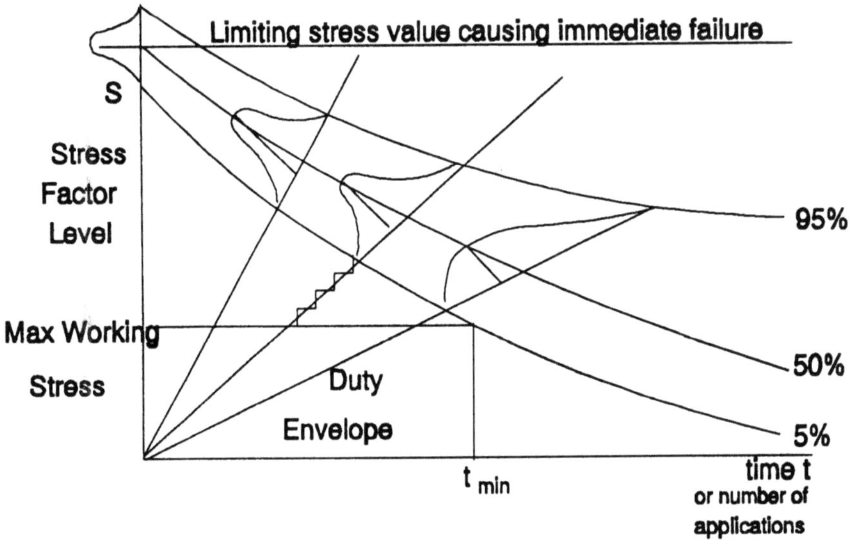

Fig. 13.10 Step-stress method of accelerated testing.

proportion failing. It is worth noting that the first, tenth, and last failures of complete sets of 19 data estimate the 5%, 50% and 95% points on the curves and that the first, fifth and last of nine estimate 10%, 50% and 90%. Possibly the quickest way to show that a duty requirement of t_{min} with, say, 95% probability of survival is to step stress three or four batches of 19 and plot the first failures. If the fair curve joining these points passes outside the duty envelope as shown on Fig. 13.10 then the component is taken to be sufficiently resistant to the stress factor. Ideally, stress would be increased continuously, but it is usual to use small, equal, increments of stress and time and to 'run in' along the working stress line until the radius is encountered, rather than starting at zero stress. Note that, as with any statistical method, the larger the samples, the better the accuracy. If large numbers of necessarily destructive tests are not feasible, then the method can still be used in a rougher way to get some idea of the likely behaviour of the component under stress. In combination with general data for the material concerned, a more limited number of tests can be used to confirm or calibrate theoretical calculations. For example, if the fatigue curve for an alloy is known, two or three step-stress failures might be enough to calibrate for the (unknown) maximum stresses developed in the component made from the alloy in terms of the forces applied to it. Appropriate radii for these calibrating readings might be found by finite-element stress analysis and Miner's rule of cumulative damage in fatigue.

(i) Data-collection analysis and feedback

In an evolutionary design the analysis should start with data arising from actual service of the present model. Pareto analysis is the simplest and probably the most effective single measure because it concentrates the minds of the (re)designers upon the major failings of the present design. In a fully revolutionary (new technology) design all the data must come from testing which must therefore be more extensive. In the case of an evolutionary design it is important to collect and analyse failure data concerning the new features and to conduct tests such that the strength (in the most general sense of resistance to failure) of redesigned components is established relative to their predecessors. Weibull analysis is really only needed if the failure mode is to be dealt with by PM rather than redesign or to determine which of several competing component designs is really best for the application.

It is important that data collection and analysis continues into the production and function and maintenance phases of the product cycle. Reliability is the time aspect of quality and companies which are not continually seeking to improve their products' quality will inevitably lose market share to companies that are and subsequently will go out of business. Data collection and analysis of customers' experiences, good and bad, is therefore vital. See also Chapter 12.

13.2.3 Production phase

As regards Reliability, this can be a danger-time. Production engineers often wish to change the design at this stage to make manufacture easier or to lower costs. The products must not be marketed without further test if initial customer complaints are to be avoided. As regards quality of conformance, there must be adequate inspection/gauging, etc., before production starts. It is no good leaving this to the last minute – it must be developed with the design. When final tooling is installed, part of the pre-production testing must be of the quality-control arrangements. Finally, the reliability of the final design made in the final way should be demonstrated.

The preservation of reliability achievement as the product moves from prototype to production is essential if the development and testing work is not to be wasted. The involvement of production personnel in the design review meetings may not be enough to prevent their suggesting last-minute changes to make the product or a component of it cheaper or easier to produce. Configuration control of the design at this stage is vital. No changes should be permitted that might affect reliability or durability without proper design investigation and testing.

(a) Quality control

The production FMECA and the testing program taken together should be used to arrive at final decisions about where QC will be applied on the production line possibly following a final series of process capability studies involving the production and assembly of components to what is hoped to be the final detailed design.

Component screening is a general term for such operations as burn-in, selective assembly, and bedding in. All these procedures are really production quality-control actions rather than reliability testing. The same applies to requirements to test the performance or safety of supplies from the production line as a routine measure. The results of these actions tell us a lot about the quality of components but less about the design's inherent reliability in service. Yet companies persist in referring to such procedures as assuring reliability. They will only do that if the design is known to be adequately reliable when built from components of good quality. Screening merely takes out weak components, the design reliability is assured by thorough testing before production starts.

Production reliability demonstration testing of models built by the final production method and sampled at random from the line may be required by customers or imposed by producers to protect their reputation.

Frequently, such testing reveals the need to adjust and tune production processes before the main run begins. It may be possible to rework the first batch to conform to requirements. Omission of this step when a long production run is planned can be fatal to the product's success. Many engineers avoid buying examples of a new model until it has had the 'wrinkles ironed out'. If the new design proves to be a success there is then a danger that QC standards will be allowed to decline in order to increase production. Attention to details like this can make the difference between a product's success or failure, or, perhaps worse, between selling thousands and turning a modest profit and selling millions and making a real killing.

13.2.4 Function and maintenance phase

Good manufacturers should be anxious to monitor the performance and R&M of their products in use. This means data collection by one means or another and a readiness to redesign unsatisfactory features. A maintenance schedule should be provided if appropriate and this, too, should be monitored, to ensure that it is adequate and not overdone. These features complete the feed-back loop for evolutionary design and are absolutely vital to its success as a design method.

Data collection, analysis and feedback should continue, using data supplied by customers or by employees using the product. These analyses can be compared with the FMECA and testing results with the object of improving future analyses and testing programs. This is called design audit.

(a) Maintenance

The FMECA and testing program will have been used to define the initial preventive schedule for the product. Using data on sales and spares consumption and directly supplied reliability data it is possible to build up a picture of the troubles being experienced by customers with the product. This analysis can be used in two ways. First, the need to redesign certain components will become apparent and can be implemented at the next design update. Second, the PM Schedule and the spare parts provision can meanwhile be adjusted to minimize the effects and costs to current customers. The methodology and mathematics of maintenance optimization are subjects for another book.

REFERENCES

BS5760 *Reliability of Systems Equipments and Components: Part 1, Guide to Reliability Programme Management*. British Standards Institution, London.

Carter, A.D.S. (1972, 1986) *Mechanical Reliability*, Macmillan, London.

Crosby, P. (1979) *Quality is Free*, McGraw-Hill, NY.

Jenney, B.W.J. (1972) *Lecture notes on Quality and Reliability*, University of Birmingham, UK.

MIL-STD-1629E (1989) *Failure Modes Effects and Criticality Analysis*, Department of Defense, Washington, DC, USA.

Miner, M.A. (1945) *J. Appl. Mech.*, **12**, *Trans ASME*, **67**, A159–64.

Tables of median, 5% and 95% ranks

Median ranks

Order no. i	Sample size n 1	2	3	4	5	6	7	8	9	10
1	50.00	29.29	20.63	15.91	12.95	10.91	9.43	8.30	7.41	6.70
2		70.71	50.00	38.57	31.38	26.45	22.85	20.11	17.96	16.23
3			79.37	61.43	50.00	42.14	36.41	32.05	28.62	25.86
4				84.09	68.62	57.86	50.00	44.02	39.31	35.51
5					87.06	73.56	63.59	55.98	50.00	45.17
6						89.09	77.15	67.95	60.69	54.83
7							90.57	79.89	71.38	64.49
8								91.70	82.04	74.14
9									92.59	83.77
10										93.30

Median ranks

Order no. i	Sample size n 11	12	13	14	15	16	17	18	19	20
1	6.11	5.61	5.19	4.83	4.52	4.24	4.00	3.78	3.58	3.41
2	14.80	13.60	12.58	11.70	10.94	10.27	9.68	9.15	8.68	8.25
3	23.58	21.67	20.05	18.65	17.43	16.37	15.42	14.58	13.83	13.15
4	32.38	29.76	27.53	25.61	23.94	22.47	21.18	20.02	18.99	18.06
5	41.19	37.85	35.02	32.58	30.45	28.59	26.94	25.47	24.15	22.97
6	50.00	45.95	42.51	39.54	36.97	34.71	32.70	30.92	29.32	27.88
7	58.81	54.05	50.00	46.52	43.48	40.82	38.47	36.37	34.49	32.80
8	67.62	62.15	57.49	53.49	50.00	46.94	44.23	41.82	39.66	37.71
9	76.42	70.24	64.98	60.46	56.52	53.06	50.00	47.27	44.83	42.63
10	85.20	78.33	72.47	67.43	63.03	59.18	55.77	52.73	50.00	47.54
11	93.89	86.40	79.96	74.39	69.55	65.30	61.53	58.18	55.17	52.46
12		94.39	87.42	81.35	76.06	71.41	67.30	63.63	60.34	57.37
13			94.81	88.30	82.57	77.53	73.06	69.08	65.51	62.29
14				95.17	89.06	83.64	78.82	74.53	70.68	67.21
15					95.48	89.73	84.58	79.98	75.85	72.12
16						95.76	90.32	85.42	81.01	77.03
17							96.01	90.85	86.17	81.95
18								96.22	91.32	86.85
19									96.42	91.75
20										96.59

Median ranks

Order no. i	*Sample size* n *21*	*22*	*23*	*24*	*25*	*26*	*27*	*28*	*29*	*30*
1	3.25	3.10	2.97	2.85	2.73	2.63	2.53	2.45	2.36	2.28
2	7.86	7.51	7.19	6.90	6.62	6.37	6.14	5.92	5.72	5.53
3	12.53	11.97	11.46	10.99	10.55	10.15	9.78	9.44	9.11	8.81
4	17.21	16.44	15.73	15.09	14.49	13.94	13.43	12.96	12.52	12.10
5	21.89	20.91	20.02	19.19	18.44	17.74	17.09	16.48	15.92	15.40
6	26.57	25.38	24.30	23.30	22.38	21.53	20.74	20.01	19.33	18.69
7	31.26	29.86	28.58	27.41	26.32	25.33	24.40	23.54	22.74	21.99
8	35.94	34.33	32.86	31.51	30.27	29.12	28.06	27.07	26.14	25.28
9	40.63	38.81	37.15	35.62	34.22	32.92	31.71	30.59	29.55	28.58
10	45.31	43.29	41.43	39.73	38.16	36.71	35.37	34.12	32.96	31.87
11	50.00	47.76	45.72	43.84	42.11	40.51	39.03	37.65	36.37	35.17
12	54.69	52.24	50.00	47.95	46.05	44.31	42.69	41.18	39.78	38.46
13	59.37	56.71	54.28	52.05	50.00	48.10	46.34	44.71	43.18	41.76
14	64.06	61.19	58.57	56.16	53.95	51.90	50.00	48.24	46.59	45.05
15	68.74	65.67	62.85	60.27	57.89	55.70	53.66	51.76	50.00	48.35
16	73.43	70.14	67.14	64.38	61.84	59.49	57.32	55.29	53.41	51.65
17	78.11	74.62	71.42	68.49	65.79	63.29	60.97	58.82	56.82	54.94
18	82.79	79.09	75.70	72.59	69.73	67.08	64.63	62.35	60.23	58.24
19	87.47	83.56	79.99	76.70	73.68	70.88	68.29	65.88	63.63	61.54
20	92.14	88.03	84.27	80.81	77.62	74.68	71.95	69.41	67.04	64.85
21	96.75	92.49	88.54	84.91	81.57	78.47	75.60	72.94	70.45	68.13
22		96.90	92.81	89.01	85.51	82.27	79.26	76.46	73.86	71.42
23			97.03	93.11	89.45	86.06	82.91	79.99	77.27	74.72
24				97.15	93.38	89.85	86.57	83.52	80.67	78.01
25					97.27	93.63	90.22	87.04	84.08	81.31
26						97.37	93.86	90.56	87.48	84.60
27							97.47	94.08	90.87	87.90
28								97.56	94.28	91.19
29									97.64	94.47
30										97.72

Median ranks

Order no. i	Sample size n 31	32	33	34	35	36	37	38	39	40
1	2.21	2.14	2.08	2.02	1.96	1.91	1.86	1.81	1.76	1.72
2	5.36	5.19	5.03	4.89	4.75	4.62	4.50	4.38	4.27	4.16
3	8.53	8.27	8.02	7.79	7.57	7.36	7.16	6.98	6.80	6.63
4	11.72	11.36	11.02	10.69	10.39	10.11	9.84	9.58	9.34	9.10
5	14.91	14.45	14.01	13.60	13.22	12.86	12.51	12.18	11.87	11.58
6	18.09	17.54	17.01	16.51	16.05	15.61	15.19	14.79	14.42	14.06
7	21.28	20.63	20.01	19.43	18.88	18.36	17.86	17.40	16.96	16.54
8	24.47	23.72	23.01	22.34	21.70	21.11	20.54	20.01	19.50	19.01
9	27.66	26.81	26.01	25.25	24.53	23.86	23.22	22.61	22.04	21.49
10	30.86	29.90	29.00	28.16	27.36	26.61	25.90	25.22	24.58	23.97
11	34.05	32.99	32.00	31.07	30.19	29.36	28.58	27.83	27.12	26.45
12	37.24	36.09	35.00	33.98	33.02	32.11	31.25	30.44	29.66	28.93
13	40.43	39.18	38.00	36.90	35.85	34.87	33.93	33.05	32.21	31.41
14	43.62	42.27	41.00	39.81	38.68	37.62	36.61	35.65	34.75	33.89
15	46.81	45.36	44.00	42.72	41.51	40.37	39.29	38.26	37.29	36.37
16	50.00	48.45	47.00	45.63	44.34	43.12	41.97	40.87	39.83	38.84
17	53.19	51.55	50.00	48.54	47.17	45.87	44.64	43.48	42.37	41.32
18	56.38	54.64	53.00	51.46	50.00	48.62	47.32	46.09	44.92	43.80
19	59.57	57.73	56.00	54.37	52.83	51.38	50.00	48.70	47.46	46.28
20	62.76	60.82	59.00	57.28	55.66	54.13	52.68	51.30	50.00	48.76
21	65.95	63.92	62.00	60.19	58.49	56.83	55.36	53.91	52.54	51.24
22	69.15	67.01	65.00	63.11	61.32	59.63	58.04	56.52	55.08	53.72
23	72.34	70.10	68.00	66.02	64.15	62.38	60.71	59.13	57.63	56.20
24	75.53	73.19	71.00	68.93	66.97	65.14	63.39	61.74	60.17	58.68
25	78.72	76.28	74.00	71.84	69.81	67.84	66.07	64.35	62.71	61.16
26	81.91	79.37	76.99	74.75	72.64	70.64	68.75	66.95	65.25	63.64
27	85.09	82.47	79.99	77.66	75.47	73.39	71.43	69.56	67.79	66.11
28	88.28	85.56	82.99	80.58	78.30	76.14	74.10	72.17	70.34	68.59
29	91.47	88.64	85.99	83.49	81.13	78.90	76.78	74.78	72.88	71.07
30	94.65	91.73	88.99	86.40	83.95	81.65	79.46	77.39	75.42	73.55
31	97.79	94.81	91.98	89.31	86.78	84.40	82.14	79.99	77.96	76.03
32		97.86	94.97	92.21	89.61	87.15	84.81	82.60	80.50	78.51
33			97.92	95.11	92.43	89.89	87.49	85.21	83.04	80.99
34				97.98	95.25	92.64	90.17	87.82	85.59	83.47
35					98.04	95.38	92.84	90.42	88.13	85.94
36						98.09	95.51	93.03	90.67	88.42
37							98.14	95.62	93.20	90.90
38								98.19	95.73	93.37
39									98.24	95.84
40										98.28

Order no. i	*Sample size* n 41	42	43	44	45	46	47	48	49	50
1	1.68	1.64	1.60	1.56	1.53	1.50	1.46	1.43	1.41	1.38
2	4.06	3.96	3.87	3.79	3.70	3.62	3.55	3.47	3.40	3.33
3	6.47	6.32	6.17	6.03	5.90	5.77	5.65	5.53	5.42	5.31
4	8.88	8.67	8.47	8.28	8.10	7.93	7.76	7.60	7.44	7.30
5	11.30	11.03	10.78	10.54	10.30	10.08	9.87	9.66	9.47	9.28
6	13.72	13.39	13.08	12.79	12.51	12.24	11.98	11.73	11.49	11.27
7	16.14	15.75	15.39	15.04	14.71	14.39	14.09	13.80	13.52	13.25
8	18.55	18.12	17.70	17.30	16.92	16.55	16.20	15.87	15.55	15.24
9	20.97	50.48	20.00	19.55	19.12	18.71	18.31	17.94	17.57	17.22
10	23.39	22.84	22.31	21.81	21.33	20.87	20.43	20.00	19.60	19.21
11	25.81	25.20	24.62	24.06	23.53	23.03	22.54	22.07	21.63	21.20
12	28.23	27.56	26.93	26.32	25.74	25.18	24.65	24.14	23.65	23.18
13	30.65	29.92	29.23	28.57	27.94	27.34	26.76	26.21	25.68	25.17
14	33.07	32.29	31.54	30.83	30.15	29.50	28.88	28.28	27.71	27.15
15	35.49	34.65	33.85	33.08	32.36	31.66	30.99	30.35	29.73	29.14
16	37.91	37.01	36.16	35.34	34.56	33.81	33.10	32.42	31.76	31.13
17	40.32	39.37	38.46	37.60	36.77	35.97	35.21	34.48	33.79	33.11
18	42.74	41.73	40.77	39.85	38.97	38.13	37.33	36.55	35.82	35.10
19	45.16	44.10	43.08	42.11	41.18	40.29	39.44	38.62	37.84	37.09
20	47.58	46.46	45.39	44.36	43.38	42.45	41.55	40.69	39.87	39.07
21	50.00	48.82	47.69	46.62	45.59	44.61	43.66	42.76	41.89	41.06
22	52.42	51.18	50.00	48.87	47.79	46.76	45.78	44.83	43.92	43.05
23	54.84	53.54	52.31	51.13	50.00	48.92	47.89	46.90	45.95	45.03
24	57.26	55.91	54.62	53.38	52.21	51.08	50.00	48.97	47.97	47.02
25	59.68	58.27	56.92	55.64	54.41	53.24	52.11	51.03	50.00	49.01
26	62.10	60.63	59.23	57.89	56.62	55.40	54.23	53.10	52.03	50.99
27	64.51	62.99	61.54	60.15	58.82	57.55	56.34	55.17	54.05	52.98
28	66.93	65.35	63.85	62.41	61.03	59.71	58.45	57.24	56.08	54.97
29	69.35	67.71	66.15	64.66	63.23	61.87	60.56	59.31	58.11	56.95
30	71.77	70.08	68.46	66.92	65.44	64.03	62.68	61.38	60.13	58.94
31	74.19	72.44	70.77	69.17	67.65	66.19	64.77	63.45	62.16	60.93
32	76.61	74.80	73.07	71.43	69.85	68.34	66.90	65.52	64.19	62.91
33	79.03	77.16	75.38	73.68	72.06	70.50	69.01	67.59	66.22	64.90
34	81.45	79.52	77.69	75.94	74.26	72.66	71.13	69.65	68.24	66.89
35	83.87	81.89	80.00	78.19	76.47	74.82	73.24	71.72	70.27	68.87
36	86.28	84.25	82.30	80.45	78.67	76.98	75.35	73.79	72.30	70.86
37	88.70	86.61	84.61	82.70	80.88	79.13	77.46	75.86	74.32	72.65
38	91.12	86.97	86.92	84.96	83.08	81.29	79.57	77.93	76.35	74.83
39	93.53	91.33	89.22	87.21	85.28	83.45	81.69	80.00	78.38	76.82
40	95.94	93.68	91.53	89.47	87.49	85.61	83.80	82.07	80.40	78.81
41	98.32	96.04	93.83	91.72	89.70	87.76	85.91	84.13	82.43	80.79
42		95.36	96.13	93.97	91.90	89.92	88.02	86.20	84.46	82.78
43			98.40	96.22	94.10	92.08	90.13	88.27	86.48	84.76
44				98.44	96.30	94.23	92.24	90.34	88.51	86.75
45					98.47	96.38	94.35	92.40	90.53	88.74
46						98.50	96.46	94.47	92.56	90.72
47							98.54	96.53	94.58	92.71
48								98.57	96.60	94.69
49									98.60	96.67
50										98.62

5% ranks

Order no. i	*Sample size* n 1	2	3	4	5	6	7	8	9	10
1	5.00	2.53	1.70	1.27	1.02	0.85	0.73	0.64	0.57	0.51
2		22.36	7.64	9.76	7.64	6.29	5.34	4.64	4.10	3.68
3			36.84	24.86	18.93	15.32	12.88	11.11	9.78	8.73
4				47.24	34.26	27.13	22.53	19.29	16.88	15.00
5					54.93	41.82	34.13	28.92	25.14	22.24
6						60.70	47.93	40.03	34.49	30.35
7							65.18	52.93	45.04	39.34
8								68.77	57.09	49.31
9									71.69	60.58
10										74.11

5% ranks

Order no. i	*Sample size* n 11	12	13	14	15	16	17	18	19	20
1	0.47	0.43	0.39	0.37	0.34	0.32	0.30	0.29	0.27	0.26
2	3.33	3.05	2.81	2.60	2.42	2.27	2.13	2.01	1.90	1.81
3	7.88	7.19	6.61	6.11	5.69	5.32	4.99	4.70	4.45	4.22
4	13.51	12.29	11.27	10.41	9.67	9.03	8.46	7.97	7.53	7.14
5	19.96	18.10	16.57	15.27	14.17	13.21	12.38	11.64	10.99	10.41
6	27.13	24.53	22.40	20.61	19.09	17.78	16.64	15.63	14.75	13.96
7	34.98	31.52	28.71	26.36	24.37	22.67	21.19	19.90	18.75	17.73
8	43.56	39.09	35.48	32.50	30.00	27.86	26.01	24.40	22.97	21.71
9	52.99	47.27	42.74	39.04	35.96	33.34	31.08	29.12	27.40	25.87
10	63.56	56.19	50.54	46.00	42.26	39.10	36.40	34.06	32.01	30.20
11	76.16	66.13	58.99	53.43	48.93	45.17	41.97	39.22	36.81	34.69
12		77.91	68.37	61.46	56.02	51.56	47.81	44.60	41.81	39.36
13			79.42	70.33	63.66	58.34	53.95	50.22	47.00	44.20
14				80.74	72.06	65.62	60.44	56.11	52.42	49.22
15					81.90	73.60	67.38	62.33	58.09	54.44
16						82.93	74.99	68.97	64.06	59.90
17							83.84	76.23	70.42	65.63
18								84.67	77.36	71.74
19									85.41	78.39
20										86.09

5% ranks

Order no. i	*Sample size* n 21	22	23	24	25	26	27	28	29	30
1	0.24	0.23	0.22	0.21	0.21	0.20	0.19	0.18	0.18	0.17
2	1.72	1.64	1.57	1.50	1.44	1.38	1.33	1.28	1.24	1.20
3	4.01	3.82	3.65	3.50	3.35	3.22	3.10	2.99	2.88	2.78
4	6.78	6.46	6.17	5.90	5.66	5.43	5.22	5.03	4.85	4.69
5	9.88	9.41	8.98	8.59	8.23	7.90	7.59	7.31	7.05	6.81
6	13.25	12.60	12.02	11.49	11.01	10.56	10.15	9.77	9.42	9.09
7	16.82	15.99	15.25	14.57	13.95	13.38	12.85	12.37	11.92	11.50
8	20.58	19.56	18.63	17.80	17.03	16.33	15.68	15.09	14.53	14.02
9	24.50	23.27	22.16	21.16	20.24	19.40	18.62	17.91	17.25	16.63
10	28.58	27.13	25.82	24.64	23.56	22.57	21.66	20.82	20.05	19.33
11	32.81	31.13	29.61	28.24	26.99	25.84	24.79	23.83	22.93	22.11
12	37.19	35.25	33.52	31.94	30.51	29.51	28.01	26.91	25.89	24.95
13	41.72	39.52	37.54	35.76	34.14	32.66	31.31	30.07	28.93	27.87
14	46.41	43.91	41.68	39.68	37.86	36.21	34.70	33.31	32.03	30.85
15	51.26	48.45	45.95	43.71	41.68	39.84	38.16	36.62	35.20	33.89
16	56.30	53.15	50.36	47.86	45.61	43.57	41.71	40.00	38.44	37.00
17	61.56	58.02	54.90	52.13	49.64	47.38	45.34	43.46	41.75	40.16
18	67.08	63.09	59.61	56.53	53.78	51.30	49.05	47.00	45.12	43.39
19	72.95	68.41	64.51	61.09	58.05	55.32	52.86	50.62	48.57	46.69
20	79.33	74.05	69.64	65.82	62.46	59.47	56.77	54.33	52.10	50.06
21	86.71	80.19	75.08	70.77	67.04	63.74	60.79	58.13	55.71	53.49
22		87.27	80.98	76.02	71.83	68.18	64.94	62.03	59.40	57.01
23			87.79	81.71	76.90	72.81	69.24	66.06	63.20	60.61
24				88.27	82.39	77.71	73.73	70.23	67.11	64.30
25					88.71	83.02	78.47	74.58	71.17	68.10
26						89.12	83.60	79.18	75.39	72.04
27							89.50	84.15	79.84	76.14
28								89.85	84.66	80.47
29									90.19	85.14
30										90.50

5% ranks

Order no. i	Sample size n 31	32	33	34	35	36	37	38	39	40
1	0.17	0.16	0.16	0.15	0.15	0.14	0.14	0.14	0.13	0.13
2	1.16	1.12	1.09	1.06	1.03	1.00	0.97	0.94	0.92	0.90
3	2.69	2.60	2.52	2.45	2.38	2.31	2.25	2.19	2.13	2.08
4	4.53	4.38	4.25	4.12	4.00	3.89	3.78	3.68	3.58	3.49
5	6.58	6.37	3.17	5.98	5.80	5.64	5.48	5.33	5.19	5.06
6	8.78	8.50	8.23	7.98	7.74	7.52	7.31	7.11	6.92	6.74
7	11.11	10.75	10.40	10.08	9.78	9.50	9.23	8.98	8.74	8.51
8	13.54	13.09	12.68	12.28	11.91	11.57	11.24	10.93	10.64	10.36
9	16.06	15.53	15.03	14.56	14.12	13.71	13.32	12.95	12.60	12.27
10	18.66	18.04	17.46	16.91	16.40	15.91	15.46	15.03	14.62	14.24
11	21.34	20.62	19.95	19.32	18.73	18.18	17.65	17.16	16.69	16.25
12	24.08	23.26	22.50	21.79	21.12	20.49	19.90	19.34	18.81	18.31
13	26.88	25.97	25.11	24.31	23.56	22.86	22.19	21.57	20.97	20.41
14	29.75	28.73	27.78	26.88	26.05	25.27	24.53	23.83	23.18	22.55
15	32.67	31.54	30.49	29.51	28.59	27.72	26.91	26.14	25.41	24.73
16	35.66	34.42	33.26	32.18	31.17	30.22	29.32	28.48	27.69	26.94
17	38.70	37.34	36.07	34.89	33.79	32.75	31.78	30.87	30.00	29.19
18	41.80	40.32	38.94	37.66	36.46	35.33	34.28	33.28	32.35	31.46
19	44.96	43.35	41.66	40.47	39.17	37.95	36.81	35.74	34.73	33.77
20	48.18	46.44	44.82	43.32	41.92	40.61	39.38	38.22	37.14	36.11
21	51.46	49.58	47.84	46.23	44.72	43.31	41.99	40.75	39.58	38.48
22	54.81	52.79	50.91	49.18	47.56	46.05	44.63	43.31	42.06	40.88
23	58.23	56.06	54.34	52.18	50.45	48.83	47.32	45.90	44.57	43.31
24	64.74	59.31	57.24	55.24	53.39	51.66	50.05	48.53	47.11	45.78
25	65.34	62.81	60.49	58.36	56.37	54.53	52.81	51.20	49.69	48.28
26	69.04	66.31	63.82	61.53	59.42	57.45	55.62	53.91	52.31	50.81
27	72.56	69.92	67.24	64.75	62.52	60.43	58.48	56.67	54.97	53.37
28	76.65	73.64	70.75	68.11	65.70	63.48	61.39	59.46	57.66	55.97
29	81.05	77.52	74.38	71.54	68.94	66.56	64.36	62.31	60.40	58.61
30	85.59	81.61	76.15	75.07	72.28	69.73	67.38	65.21	63.19	61.29
31	90.79	86.02	82.13	78.75	75.73	72.99	70.48	68.17	66.02	64.02
32		91.06	86.42	82.62	79.31	76.35	73.66	71.20	68.92	66.80
33			91.32	86.79	83.09	79.85	76.95	74.30	71.88	69.63
34				91.57	87.15	83.53	80.36	77.51	74.92	72.53
35					91.80	87.49	83.95	80.84	78.05	75.50
36						92.02	87.81	84.34	81.30	78.56
37							92.22	88.12	84.72	81.74
38								92.42	88.41	85.09
39									92.61	88.68
40										92.78

5% ranks

Order no. i	*Sample size* n *41*	*42*	*43*	*44*	*45*	*46*	*47*	*48*	*49*	*50*
1	0.13	0.12	0.12	0.12	0.11	0.11	0.11	0.11	0.11	0.10
2	0.87	0.85	0.83	0.81	0.80	0.78	0.76	0.75	0.73	0.72
3	2.02	1.98	1.93	1.88	1.84	1.80	1.76	1.73	1.69	1.66
4	3.40	3.32	3.24	3.17	3.09	3.03	2.96	2.90	2.84	2.78
5	4.93	4.81	4.70	4.59	4.48	4.38	4.29	4.20	4.11	4.02
6	6.57	6.41	6.26	6.11	5.97	5.84	5.71	5.59	5.47	5.36
7	8.30	8.09	7.90	7.71	7.54	7.37	7.21	7.05	6.90	6.76
8	10.10	9.85	9.61	9.38	9.17	8.96	8.76	8.57	8.39	8.22
9	11.96	11.66	11.38	11.11	10.85	10.61	10.37	10.15	9.93	9.73
10	13.87	13.53	13.20	12.88	12.58	12.30	12.02	11.76	11.51	11.27
11	15.83	15.44	15.06	14.70	14.36	14.03	13.72	13.42	13.13	12.86
12	17.84	17.39	16.96	16.55	16.17	15.80	15.44	15.11	14.78	14.47
13	19.88	19.38	18.90	18.45	18.01	17.60	17.20	16.83	16.46	16.12
14	21.96	21.41	20.88	20.37	19.89	19.43	18.99	18.57	18.17	17.79
15	24.08	23.47	22.88	22.33	21.80	21.29	20.81	20.35	19.91	19.49
16	26.23	25.56	24.92	24.31	23.73	23.18	22.65	22.15	21.67	21.21
17	28.41	27.68	26.98	26.32	25.69	25.10	24.52	23.98	23.46	22.96
18	30.62	29.83	29.08	28.36	27.68	27.03	26.42	25.83	25.26	24.72
19	32.87	32.01	31.20	30.43	29.70	29.00	28.33	27.70	27.09	26.51
20	35.14	34.22	33.35	32.52	31.73	30.98	30.27	29.59	28.94	28.31
21	37.44	36.46	35.52	34.64	33.79	32.99	32.23	31.50	30.80	30.14
22	39.77	38.72	37.72	36.78	35.88	35.03	34.21	33.43	32.69	31.98
23	42.13	41.01	39.95	38.94	37.99	37.08	36.21	35.39	34.60	33.85
24	44.52	43.33	42.20	41.13	40.12	39.15	38.24	37.36	36.52	35.73
25	46.94	45.67	44.48	43.35	42.27	41.25	40.28	39.35	38.47	37.63
26	49.39	48.05	46.79	45.59	44.45	43.37	42.34	41.37	10.43	39.54
27	51.87	50.45	49.12	47.85	46.65	45.51	44.43	43.40	42.42	41.48
28	54.39	52.89	51.48	50.14	48.88	47.68	46.54	45.45	44.42	43.43
29	56.94	55.36	53.87	52.46	51.13	49.87	48.67	47.52	46.44	45.40
30	59.52	57.86	56.29	54.81	53.41	52.08	50.82	49.62	48.48	47.39
31	62.15	60.39	58.74	57.18	55.71	54.32	52.99	51.73	50.54	49.40
32	64.82	62.97	61.23	59.59	58.04	56.58	55.19	53.87	52.62	51.42
33	67.54	56.59	63.75	62.03	60.40	58.87	57.41	56.03	54.72	53.47
34	70.31	68.25	66.32	64.51	62.80	61.19	59.66	58.22	56.84	55.54
35	73.15	70.96	68.93	67.02	65.23	63.54	61.94	60.43	58.99	57.63
36	76.05	73.74	71.59	69.58	67.69	65.92	64.25	62.66	61.16	59.74
37	79.05	76.58	74.31	72.19	70.20	68.34	66.59	64.93	63.36	61.87
38	82.16	79.52	77.09	74.85	72.76	70.81	68.96	67.23	65.59	64.03
39	85.43	82.56	79.96	77.58	75.37	73.31	71.38	69.56	67.85	66.22
40	88.95	85.76	92.94	80.39	78.05	75.87	73.84	71.93	70.14	68.44
41	92.95	89.20	86.07	83.31	80.80	78.49	76.35	74.35	72.47	70.69
42		93.12	89.44	86.37	83.66	81.20	78.92	76.81	74.84	72.98
43			93.27	89.67	86.66	84.00	81.57	79.34	77.26	75.31
44				93.42	89.89	86.94	84.32	81.94	79.73	77.68
45					93.56	90.10	87.20	84.63	82.29	80.12
46						93.70	90.30	87.46	84.93	82.62
47							93.83	90.49	87.70	85.22
48								93.95	90.68	87.94
49									94.07	90.86
50										94.18

95% ranks

Order no. i	*Sample size* n *1*	*2*	*3*	*4*	*5*	*6*	*7*	*8*	*9*	*10*
1	95.00	77.64	63.16	52.71	45.07	39.30	34.82	31.23	28.31	25.89
2		97.47	86.47	75.14	65.74	58.18	52.07	47.07	42.91	39.42
3			98.31	90.24	81.08	72.87	65.87	59.97	54.96	50.69
4				98.73	92.36	84.68	77.47	71.08	65.51	60.66
5					98.98	93.72	87.12	80.71	74.86	69.65
6						99.15	94.66	88.89	83.13	77.76
7							99.27	95.36	90.23	85.00
8								99.36	95.90	91.27
9									99.43	96.32
10										99.49

95% ranks

Order no. i	*Sample size* n *11*	*12*	*13*	*14*	*15*	*16*	*17*	*18*	*19*	*20*
1	23.84	22.09	20.58	19.26	18.10	17.08	16.16	15.33	14.59	13.91
2	36.44	33.87	31.63	29.67	27.94	26.40	25.01	23.77	22.64	21.61
3	47.01	43.81	41.01	38.54	36.34	34.38	32.62	31.03	29.58	28.26
4	56.44	52.73	49.47	46.57	43.98	41.66	39.56	37.67	35.94	34.37
5	65.02	60.91	57.26	54.00	51.08	48.44	46.06	43.89	41.91	40.10
6	72.88	68.48	64.52	60.93	57.74	54.84	52.19	49.78	47.58	45.56
7	80.04	75.47	71.30	67.50	64.04	60.90	58.03	55.40	53.00	50.78
8	86.49	81.90	77.60	73.64	70.00	66.66	63.60	60.78	58.19	55.80
9	92.12	87.72	83.43	79.39	75.63	72.14	68.92	65.94	63.19	60.64
10	96.67	92.81	88.73	84.73	80.91	77.33	73.99	70.88	67.99	65.31
11	99.54	96.95	93.40	89.60	85.83	82.22	78.81	75.60	72.61	69.81
12		99.57	97.20	93.89	90.33	86.79	83.36	80.11	77.03	74.14
13			99.61	97.40	94.32	90.98	87.62	84.37	81.25	78.29
14				99.63	97.58	94.69	91.54	88.36	85.25	82.27
15					99.66	97.73	95.01	92.03	89.01	86.05
16						99.68	97.87	95.30	92.47	89.59
17							99.70	97.99	95.55	92.87
18								99.72	98.10	95.78
19									99.73	98.19
20										99.74

95% ranks

Order no. i	*Sample size* n 21	22	23	24	25	26	27	28	29	30
1	13.30	12.73	12.21	11.74	11.29	10.88	10.50	10.15	9.81	9.50
2	20.67	19.81	19.02	18.29	17.61	16.98	16.40	15.85	15.34	14.86
3	27.06	25.95	24.93	23.98	23.10	22.29	21.53	20.82	20.16	19.53
4	32.92	31.59	30.36	29.23	28.17	27.19	26.27	25.42	24.61	23.86
5	38.44	36.91	35.19	34.18	32.96	31.82	30.76	29.77	28.84	27.96
6	43.70	41.98	40.39	38.91	37.54	36.26	35.06	33.94	32.89	31.90
7	48.74	46.85	45.10	43.47	41.95	40.54	39.21	37.97	36.80	35.70
8	53.59	51.55	49.64	47.87	46.22	44.68	43.23	41.87	40.60	39.40
9	58.28	56.09	54.05	52.14	50.36	48.70	47.14	45.67	44.29	42.99
10	62.81	60.48	58.32	56.29	54.39	52.62	50.95	49.38	47.90	46.51
11	67.19	64.75	62.46	60.32	58.32	56.43	54.66	53.00	51.43	49.94
12	71.42	68.87	66.49	64.24	62.14	60.16	58.29	56.54	54.88	53.31
13	75.50	72.87	70.39	68.06	65.86	63.79	61.84	60.00	58.25	56.61
14	79.43	76.73	74.18	71.76	69.49	67.34	65.30	63.38	61.56	59.84
15	83.18	80.44	77.84	75.36	73.02	70.79	68.69	66.69	64.80	63.01
16	86.76	84.01	81.37	78.84	76.44	74.16	71.99	69.93	67.97	66.11
17	90.12	87.40	84.75	82.20	79.76	77.43	75.21	73.09	71.07	69.15
18	93.22	90.59	87.98	85.43	82.97	80.60	78.34	76.17	74.11	72.13
19	95.99	93.54	91.02	88.51	86.05	83.67	81.38	79.18	77.07	75.05
20	98.28	96.18	93.83	91.41	88.99	86.62	84.32	82.09	79.95	77.89
21	99.76	98.36	96.35	94.10	91.77	89.44	87.15	84.92	82.75	80.67
22		99.77	98.43	96.51	94.34	92.10	89.85	87.63	85.47	83.37
23			99.78	98.50	96.65	94.57	92.41	90.23	88.08	85.98
24				99.79	98.56	96.78	94.78	92.69	90.58	88.50
25					99.80	98.62	96.90	94.97	92.95	90.91
26						99.80	98.67	97.02	95.15	93.19
27							99.81	98.72	97.12	95.31
28								99.82	98.76	97.22
29									99.82	98.80
30										99.83

95% ranks

Order *no.* i	*Sample size* n *31*	*32*	*33*	*34*	*35*	*36*	*37*	*38*	*39*	*40*
1	9.21	8.94	8.68	8.43	8.20	7.99	7.78	7.58	7.39	7.22
2	14.41	13.99	13.59	13.21	12.85	12.51	12.19	11.89	11.60	11.32
3	18.96	18.39	17.87	17.38	16.92	16.47	16.05	15.66	15.28	14.92
4	23.15	22.48	21.85	21.25	20.69	20.15	19.64	19.16	18.70	18.26
5	27.14	26.36	25.63	24.93	24.27	23.65	23.05	22.49	21.95	21.44
6	30.96	30.08	29.25	28.47	27.72	27.01	26.34	25.70	25.09	24.50
7	34.67	33.69	32.76	31.89	31.06	30.27	29.52	28.80	28.12	27.48
8	38.26	37.19	36.18	35.22	34.31	33.44	32.62	31.83	31.08	30.37
9	41.77	40.61	39.51	68.47	37.48	36.54	35.64	34.79	33.98	33.20
10	45.19	43.95	42.77	41.65	40.58	39.57	38.61	37.69	36.82	35.98
11	48.54	57.21	45.96	44.76	43.63	42.55	41.52	40.54	39.60	38.71
12	51.83	50.42	49.09	47.82	46.62	45.47	44.38	43.33	42.34	41.39
13	55.04	53.56	52.16	50.82	49.55	48.34	47.19	46.09	45.03	44.03
14	58.20	56.65	55.18	53.78	52.44	51.17	49.96	48.80	47.69	46.63
15	61.30	59.68	58.14	56.68	55.28	53.95	52.68	51.47	50.31	49.20
16	64.34	62.66	61.06	59.53	58.08	56.69	55.37	54.10	52.89	51.73
17	67.33	65.59	63.93	62.34	60.83	59.39	58.01	56.69	55.43	54.22
18	70.25	68.46	66.74	65.11	63.54	62.05	60.62	59.25	57.94	56.69
19	73.12	71.27	69.51	67.82	66.21	64.67	63.19	61.78	60.42	59.12
20	75.92	74.03	72.23	70.49	68.84	67.25	65.72	64.26	62.86	61.52
21	78.66	76.75	74.89	73.12	71.42	69.78	68.22	66.72	65.28	63.89
22	81.34	79.38	77.50	75.69	73.95	72.28	70.68	69.14	67.65	66.23
23	83.94	81.96	80.05	78.21	76.44	74.74	73.09	71.52	70.00	68.54
24	86.46	84.47	82.55	80.68	78.88	77.15	75.47	73.86	72.31	70.82
25	88.89	86.91	84.97	83.09	81.27	79.51	77.81	76.17	74.59	73.06
26	91.22	89.26	87.33	85.44	83.60	81.83	80.10	78.44	76.83	75.27
27	93.42	91.51	89.60	87.72	85.88	84.09	82.35	80.66	79.03	77.45
28	95.47	93.64	91.77	89.92	88.09	86.29	84.54	82.84	81.19	79.59
29	97.31	95.62	93.83	92.02	90.22	88.43	86.68	84.97	83.31	81.69
30	98.84	97.40	95.75	94.02	92.26	90.50	88.76	87.05	85.38	83.75
31	99.84	98.88	97.48	95.88	94.20	92.48	90.77	89.07	87.40	85.76
32		99.84	98.91	97.55	96.00	94.36	92.69	91.02	89.36	87.73
33			99.85	98.95	97.62	96.11	94.52	92.89	91.26	89.64
34				99.85	98.98	97.69	96.22	94.67	93.08	91.49
35					99.85	99.00	97.75	96.32	94.81	93.26
36						99.86	99.03	97.81	96.42	94.94
37							99.86	99.06	97.87	96.51
38								99.87	99.08	97.93
39									99.87	99.10
40										99.87

95% ranks

Order no. i	Sample size n 41	42	43	44	45	46	47	48	49	50
1	7.05	6.89	6.73	6.58	6.44	6.31	6.18	6.05	5.93	5.82
2	11.06	10.80	10.56	10.33	10.11	9.90	9.70	9.51	9.32	9.14
3	14.57	14.24	13.93	13.63	13.34	13.06	12.80	12.54	12.30	12.06
4	17.84	17.44	17.06	16.69	16.34	16.00	15.68	15.37	15.07	14.79
5	20.95	20.48	20.04	19.61	19.20	18.80	18.43	18.06	17.72	17.38
6	23.95	23.42	22.91	22.42	21.95	21.51	21.08	20.66	20.27	19.88
7	26.85	26.26	25.69	25.15	24.63	24.13	23.65	23.19	22.74	22.32
8	29.69	29.04	28.41	27.81	27.24	26.69	26.16	25.62	25.16	24.69
9	32.46	31.75	31.07	30.42	29.80	29.20	28.62	28.07	27.54	27.02
10	35.18	34.42	33.68	32.98	32.31	31.66	31.04	30.44	29.86	29.31
11	37.85	37.03	36.25	35.50	34.77	34.08	33.41	32.77	32.15	31.56
12	40.48	39.61	38.77	37.97	37.20	36.46	35.75	35.07	34.41	33.78
13	43.07	42.14	41.26	40.41	39.60	38.81	38.06	37.34	36.70	35.93
14	45.62	44.64	43.71	42.82	41.96	41.13	40.34	39.57	38.84	38.13
15	48.13	47.11	46.13	45.19	44.29	43.42	42.59	41.78	41.01	40.26
16	50.61	49.55	48.52	47.54	46.59	45.67	44.81	43.97	43.16	42.37
17	53.06	51.95	50.88	49.86	48.87	47.92	47.01	46.13	45.28	44.46
18	55.48	54.33	53.22	52.15	51.12	50.13	49.18	48.27	47.38	46.53
19	57.87	56.67	55.52	54.41	53.35	52.32	51.33	50.38	49.46	48.58
20	60.23	58.99	57.80	56.65	55.55	54.49	53.46	52.48	51.52	50.60
21	62.56	61.28	60.05	58.87	57.73	56.63	55.57	54.55	53.56	52.61
22	64.86	63.55	62.27	61.06	59.88	58.75	57.56	56.60	55.58	54.80
23	67.13	65.78	64.48	63.22	62.01	60.85	59.72	58.63	57.59	56.57
24	69.38	67.99	66.65	65.36	64.12	62.92	61.77	60.65	59.57	58.52
25	71.59	70.17	68.80	67.48	66.21	64.98	63.79	62.64	61.53	60.46
26	73.77	72.32	70.92	69.57	68.27	67.01	65.79	64.61	63.48	62.38
27	75.92	74.44	73.02	71.64	70.30	69.02	67.77	66.57	65.40	64.27
28	78.04	76.53	75.08	73.68	72.32	71.00	69.73	68.50	67.31	66.16
29	80.12	78.59	77.12	75.69	74.31	72.97	71.67	70.41	69.20	68.02
30	82.16	80.62	79.13	77.67	76.27	74.91	73.58	72.30	71.06	69.86
31	84.17	82.61	81.10	79.63	78.20	76.82	75.48	74.18	72.91	71.69
32	86.13	84.56	83.04	81.55	80.11	78.71	77.35	76.02	74.74	73.49
33	88.04	86.48	84.94	83.45	81.99	80.57	79.19	77.85	76.55	75.28
34	89.90	88.34	86.81	85.30	83.83	82.40	81.01	79.65	78.33	77.05
35	91.70	90.15	88.62	87.12	85.65	84.20	82.80	81.43	80.09	78.79
36	93.43	91.91	90.39	88.89	87.42	85.97	84.56	83.18	81.83	80.51
37	95.07	93.59	92.10	90.62	89.15	87.70	86.29	94.90	83.54	82.21
38	96.60	95.19	93.74	92.29	90.83	89.40	87.98	86.59	85.22	83.88
39	97.98	96.68	95.31	93.89	92.46	91.04	89.63	88.24	86.87	85.53
40	99.13	98.03	96.76	95.41	94.03	92.63	91.24	89.85	88.49	87.14
41	99.88	99.15	98.07	96.84	95.52	94.16	92.80	91.43	90.07	88.73
42		99.88	99.17	98.12	96.91	95.62	94.29	92.95	91.61	90.28
43			99.88	99.19	98.16	96.98	95.71	94.41	93.10	91.78
44				99.88	99.21	98.20	97.04	95.81	94.53	93.24
45					99.89	99.22	98.24	97.10	95.89	94.64
46						99.89	99.24	98.28	97.16	95.98
47							99.89	99.26	98.31	97.22
48								99.89	99.27	98.35
49									99.90	99.29
50										99.90

APPENDIX B

Gamma function $\int_0^{\infty} e^{-x} x^{n-1} dx$

The table is calculated from the Stirling approximation given in Chapter 12.

n	Γn	n	Γn	n	Γn	n	Γn
1.00	1.00000	1.25	0.90640	1.50	0.88623	1.75	0.91906
1.01	0.99433	1.26	0.90440	1.51	0.88659	1.76	0.92137
1.02	0.98884	1.27	0.90250	1.52	0.88704	1.77	0.92376
1.03	0.98355	1.28	0.90072	1.53	0.88757	1.78	0.92623
1.04	0.97844	1.29	0.89904	1.54	0.88818	1.79	0.92877
1.05	0.97350	1.30	0.89747	1.55	0.88887	1.80	0.93138
1.06	0.96874	1.31	0.89600	1.56	0.88964	1.81	0.93408
1.07	0.96415	1.32	0.89464	1.57	0.89049	1.82	0.93685
1.08	0.95973	1.33	0.89338	1.58	0.89142	1.83	0.93969
1.09	0.95546	1.34	0.89222	1.59	0.89243	1.84	0.94261
1.10	0.95135	1.35	0.89115	1.60	0.89352	1.85	0.94561
1.11	0.94739	1.36	0.89018	1.61	0.89468	1.86	0.94869
1.12	0.94359	1.37	0.88931	1.62	0.89592	1.87	0.95184
1.13	0.93993	1.38	0.88854	1.63	0.89724	1.88	0.95507
1.14	0.93642	1.39	0.88785	1.64	0.89864	1.89	0.95838
1.15	0.93304	1.40	0.88726	1.65	0.90012	1.90	0.96177
1.16	0.92980	1.41	0.88676	1.66	0.90167	1.91	0.96523
1.17	0.92670	1.42	0.88636	1.67	0.90330	1.92	0.96878
1.18	0.92373	1.43	0.88604	1.68	0.90500	1.93	0.97240
1.19	0.92088	1.44	0.98580	1.69	0.90678	1.94	0.97610
1.20	0.91817	1.45	0.88565	1.70	0.90864	1.95	0.97988
1.21	0.91558	1.46	0.88560	1.71	0.91057	1.96	0.99374
1.22	0.91311	1.47	0.88563	1.72	0.91258	1.97	0.98768
1.23	0.91075	1.48	0.88575	1.73	0.91466	1.98	0.99171
1.24	0.90852	1.49	0.88593	1.74	0.91683	1.99	0.99581
						2.00	1.00000

A2 PROCEDURE

First reduce n to a number between 1 and 2 by using $\Gamma(n + 1) = n\Gamma n$ and then enter table.

Note that for integer n, $\Gamma n = (n - 1)!$

List of symbols and abbreviations

A	Steady-state availability, MTT(B)F/[MTT(B)F + MTTR].
a	Unit availability $\mu/(\mu + \lambda) = \theta/(\theta + \phi)$
A_i	Availability of stage i.
A_{sys}	System availability (steady state unless followed by t).
$A(t)$	Probability item is available at time t, given available at time 0, called time-dependent availability; instantaneous availability, $A(O) = 1$ usually implied.
$\bar{A}(t)$	Average availability over a period 0 to t.
$A \cap B$	Intersection of events A and B (both).
$A \cup B$	Union of events A and B (either or both).
$B(v,w)$	Beta function with parameters $v, w, \Gamma v \Gamma w / \Gamma(v + w)$.
c	Shape parameter gamma distribution, cost per unit time.
C_m	Criticality number in MIL-STD-1629.
E_i	Productiveness of stage i.
F, $\bar{F}$	Number of failures, mean number of failures.
$F(t)$	Cumulative distribution function of $f(t)$ (c.d.f.).
$f(t)$	Probability density function of ages at failure of a component (p.d.f.).
$f_{\bar{T}}$	Exit frequency of top event.
f_T	Occurrence frequency of top event.
$f(\bar{\phi}_i)$	Exit frequency of system state Φ_i.
$f(\phi_i)$	Enter frequency of system state Φ_i.
$f(\phi_{ij})$	Frequency of transition from system state Φ_i to system state Φ_j.
G	All terminals of graph G connected (not connected).
$g(t_r)$	p.d.f. of repair times.
$G_{x,y}$	Terminals x and y of graph G connected.
HPP	Homogeneous Poisson process.
$H(t)$	Cumulative hazard function, integral of $z(t)$ from 0 to t.
I	Identity matrix.
i.i.d.	Identically and independently distributed (of random variables).
i, j, k	Index numbers, e.g. $A_{i,j}$ = availability of jth item in stage i.

Lt	Limit as t tends to infinity, $t \to \infty$.
m	Number of items required in a redundant system or stage.
M_i	Mean order number of the ith failure.
$\text{Min}[x,y]$	Minimum of alternatives x, y (Max. similar).
$\underset{x}{\text{Min}}[f(x)]$	Minimum value of $f(x)$ over all values of x (Max. similar).
MTBF	Mean time between failures.
MTTF	Mean time to failure (of item which fails only once).
MTTFF	Mean time to first failure (from new usually implied).
MTTR	Mean time to repair.
$M(t_r)$	Maintainability, c.d.f. of repair times; maintainability to time t_r (probability that repair is completed within t_r).
n	Number of trials, number tested, number of data, number of items in a stage.
N	Number of series stages in a system.
NHPP	Non-homogeneous Poisson process.
ϕ	Mean of $g(t)$, mean time to repair (MTTR).
P_A	Probability of acceptance in a sampling test.
$P(B\|A)$	Probability of event B given event A has occurred.
p.d.f.	Probability density function.
$p_{i,j}$	Steady-state probability of state j in stage i.
$p_{i,j}(t)$	Probability of state j in stage i at time t.
$P(X)$	Probability of event X.
Q	Matrix of conditional transition rates.
$Q(G)$	All-terminal unreliability of graph G.
$Q(G_{x,y})$	Two-terminal unreliability of graph G.
$q_{i,j}$	Element of Q in ith row and jth column.
r	Number of repairmen; rejection number in sampling plan.
RBD	Reliability block diagram.
$R(G)$	All-terminal reliability of graph G.
$R(G_{x,y})$	Two-terminal reliability of graph G.
ROCOF	Rate of occurrence of failures of a repairable system.
$R(t)$	Reliability function, $1 - F(t)$.
S	Slope in a Wald sequential test plan.
s.d.	Standard deviation (square root of second moment of area of p.d.f. about the mean).
S_i	Number of survivors just before the ith failure.
S_1, S_2, etc.	States of a system in order of preference.
T	Age of a system containing fallible components; top event of a fault tree.
t	Age at failure of a single component; time variable for components.
t_r	Time to repair a single component or mode of failure.
u	Dummy variable of integration.
$v_{i,j}$	Output of stage i in state j relative to system maximum.

$w_{k,h}, W$	Sum of probabilities of h greatest production rates of stage k, vector of Ws over all stages of system.
z	Scale parameter gamma distribution; hazard.
$z(t)$	Hazard rate function $f(t)/R(t)$.
α	Slope of Duane plot; producer's risk.
β	Shape parameter of Weibull distribution; consumer's risk.
Γ	Gamma function.
η	Scale parameter (characteristic life) of Weibull distribution.
θ	Mean of $f(t)$; Mean time to failure, MTTF of component or MTBF of a system.
λ	Constant hazard rate (exponential distribution) or constant rate of occurrence of failures (ROCOF) for a system.
$\lambda(T)$	ROCOF in a repairable system, which varies with system age T.
μ	Repair rate, $1/\phi$, constant rate usually implied.
Π	Product.
ρ	Ratio of MTBF to MTTR, $\theta/\phi = \mu/\lambda = a/(1 - a)$.
σ	Standard deviation.
ϕ_{ij}	Transition from system state Φ_i to Φ_j.
Φ_i	System state i.
$\Phi(x)$	Area under standard normal curve from $-\infty$ to mean plus x s.d.s.
Σ	Summation.
$\oplus$	Kronecker sum.
$\otimes$	Kronecker product.

Index

MIX
Papier aus verantwortungsvollen Quellen
Paper from responsible sources
FSC® C105338

If you have any concerns about our products,
you can contact us on
ProductSafety@springernature.com

In case Publisher is established outside the EU,
the EU authorized representative is:
Springer Nature Customer Service Center GmbH
Europaplatz 3, 69115 Heidelberg, Germany

Printed by Libri Plureos GmbH
in Hamburg, Germany